Microstructured Polymer Optical Fibres

Maryanne C. J. Large · Leon Poladian
Geoff W. Barton · Martijn A. van Eijkelenborg

Microstructured Polymer Optical Fibres

 Springer

Maryanne C. J. Large
University of Sydney
School of Physics
Sydney NSW 2006
Australia

Leon Poladian
University of Sydney
School of Mathematics and Statistics
Sydney NSW 2006
Australia

Geoff W. Barton
University of Sydney
Department of Chemical
  and Biomolecular Engineering
Sydney NSW 2006
Australia

Martijn A. van Eijkelenborg
University of Sydney
Optical Fibre Technology Centre
206 National Innovation Centre
Eveleigh NSW 1430
Australia

Library of Congress Control Number: 2007930766

ISBN 978-0-387-31273-6   e-ISBN 978-0-387-68617-2

9 8 7 6 5 4 3 2 1

springer.com

The work presented in this book is the culmination of six years work at the Optical Fibre Technology Centre, where all of the authors have been based at one time or another. While most of us have ended up in other departments, we would like to acknowledge the role the OFTC played in making this work possible. It has been an extraordinary place to work.

Preface

This book is intended to provide a concise and accessible introduction to microstructured polymer optical fibres (mPOF). Authors of books in any technical field need to make decisions about just how technical the explanations should be, and we have taken the view that this book should be accessible not only to academics but also, for example, to engineers who may wonder if mPOF have anything to offer for their applications, or interested undergraduate students. We have therefore aimed at conveying a correct conceptual understanding of the ideas in the clearest possible way. Even within the theoretical sections of the book we have tried to avoid using equations as the explanation of first resort. Indeed, often the same idea is explained in multiple ways, first at a very conceptual level and then later in the book in more detail. However, correctness is never sacrificed to achieve simplicity. Fuller and more technical explanations can be found in the referenced scientific literature, and or in one of the several other books available in the general area of microstructured fibres.

Much of the material in this book applies to microstructured fibres made from any material, such as the explanations of how the fibres work, much of the fabrication, the modelling techniques and some of the characterisation techniques and applications. However the book is primarily about mPOF. The community working in silica microstructured fibres (more commonly known as "Photonic crystal fibres" or PCF) is much more extensive than that working in polymers, and many of the applications they have explored have not been attempted in mPOF, and in some cases are unsuitable for them. But while mPOF is a younger technology, it also extends the scope of microstructured fibres in important ways. The large suite of fabrication techniques available in polymers mean not only that the fibres can be mass produced, but also make it much easier to produce different hole structures. In silica, most microstructured fibres use the "stack and draw" technique, where a preform is constructed by stacking capillaries and rods, which is then drawn to fibre. This technique, while versatile in many ways, restricts the types of structures that can be made. By contrast, mPOFs can and have been made with a wide variety of hole arrangements. In addition, many of the applications that are targeted in polymers require thick multimode fibres, often with extremely large cores, while PCFs are usually single or few-moded and are no longer flexible at large diameters. Finally, the material properties of polymer are very different to those of silica, and they can be modified in many more ways, not only by doping, but also by, for example, co-polymerisation or the attachment of other active groups. The area of material modification (the topic of Chapter 11) has not been treated extensively in the book primarily because little work has been done in this area. One of our intentions in writing this book is to stimulate interest in the field, and a greater involvement of chemists and material scientists would be very welcome, though we add as a caveat that any new materials used should remain drawable and preferably also highly transparent.

The book is conceptually, though not formally divided into two parts. We have tried to make each chapter relatively self contained, to make it easier to extract the information quickly without extensive cross-referencing.

In the first part of the book we introduce the ideas behind microstructured fibres. A history and description of both polymer fibres (POFs) and microstructured optical fibres (MOFs) is given in Chapter 1. Chapter 2 focusses on the basic concepts of waveguide theory, while Chapter 3, shows how the properties of microstructured fibres differ from those of conventional fibres. One of the challenges posed by the wealth of new and potentially complex designs enabled by microstructured fibres has been to be able to understand the effect of the fibre geometry on the optical properties. The modelling of microstructured fibres in Chapter 4 compares the mathematical approaches that have been used to model microstructured fibres, and extends them to include approaches to design. The range of fabrication methods available to mPOF fabrication is presented in Chapter 5 along with a discussion of the control of hole deformation in Chapter 6. The first part of the book concludes with a chapter on how to prepare mPOF for experiments and an outline of the most common characterisation techniques in Chapter 7.

In the second part of the book we describe a number of applications that use mPOF. The applications we have chosen are not comprehensive. We have restricted ourselves to those where the results are more than preliminary, or where the applications illustrate important new capabilities of the fibres. In particular, we aim to show how the physical principles explained in the first part of the book can be made technologically useful. Where possible we have also tried to bench-mark the performance against conventional polymer fibres where appropriate. The first two applications use microstructures by themselves to achieve their effects, while the last two applications describe modifications to the fibre by post-processing or doping.

Chapter 8 presents work on Hollow-Core mPOF (HC-mPOF), which is particularly significant as it allows guidance of wavelengths where the polymer is not considered transparent and in addition it provides a new route to beating the lowest-loss record for conventional polymer fibre. Chapter 9, which presents a Graded Index Microstructured Polymer Optical Fibre for large-core high-bandwidth FTTH applications. In Chapter 10 the fabrication and characterisation of Bragg and Long Period Gratings in mPOF is described. Chapter 11 outlines various doping methods that specifically use the hole structure to introduce organic dyes and nanoparticles into mPOFs.

The area of microstructured polymer optical fibres is developing very rapidly, and many of the results presented here will date quickly. While the authors optimistically anticipate many further editions of this book, the delay implicit in any publication can be considerable. To address this, we have established a website where the latest results will be posted, together with additional material such as animations, publications for downloading and relevant news items. The web address is: www.mpof.net.au.

Acknowledgments

The work presented in this book originates from many different mPOF research projects over the last six years and material is included that was the result of working with research students and collaborators, both Australian and international.

Firstly, and most importantly, we would like to express our sincere thanks to Joseph Zagari, Alex Argyros, Steve Manos and Nader Issa for their work as part of their postgraduate research at the Optical Fibre Technology Centre (OFTC) of the University of Sydney. They contributed a large fraction of the results presented throughout this book and without their contributions and enthusiasm this book would have been very much thinner. We also thank our current postgraduate students: Felicity Cox, Richard Lwin and Helmut Yu who are so critical to the ongoing work of our group. Mark Hiscocks, Matt Fellew, Trungta Keawfanapadol and Hong Nguyen also made contributions during the course of projects at the OFTC.

Particular thanks are due to Barry Reed, who has made almost all of our fibre preforms over the last six years, as well as many custom-designed pieces of equipment. His attention to detail, and his pride in his work are very much appreciated.

Steve Manos prepared the vast majority of the figures for the book, a task that involved running numerous simulations as well as doing the graphics. A number of other figures were adapted from publications by Nader Issa, Alex Argyros and Richard Lwin. We also thank Richard Lwin, Helmut Yu, and the Australian Key Centre for Microscopy and Microanalysis at the University of Sydney for the electron microscope images and technical assistance.

More generally, we would like to thank our colleagues at the OFTC. Ian Bassett, Geoff Henry, Sue Law and Whayne Padden contributed directly to work presented here, both as colleagues and as friends. Simon Fleming, Ron Bailey and Peter Henry provided advice and support at critical times. Shicheng Xue and Roger Tanner from the the School of Aerospace, Mechanical and Mechatronic Engineering at the University of Sydney have been long-term colleagues, and their insights into the rheology of the draw process have been extremely important. We have also benefited from the expertise of Ross McPhedran, Martijn de Sterke and Nicolae Nicorovici, at the School of Physics, particularly during the early stages of our work.

We would like to thank Christophe Barbé and Kim Finnie of the Materials Division at the Australian Nuclear Science and Technology Organisation (ANSTO) for advice, discussions and instructions on the fabrication of nanoparticles.

We also thank Francois Ladouceur at the University of New South Wales for his contributions to the work on nanoparticles and in establishing the mPOF draw tower facility. He has been a stimulating collaborator in various roles since the beginning of the mPOF development.

Tanya Monro and Heike Ebendorff-Heidepriem at the University of Adelaide were responsible for developing billet extrusion for mPOF preforms, and their expertise in this area was extremely helpful.

Karl-Friedrich Klein of the University of Applied Sciences of Friedberg, and Alexander Bachmann and Hans Poisel of the Polymer Optical Fiber Application Centre, both in Germany, have characterised many of our fibres. They could not have been more helpful in trying to unravel the behaviour of the high bandwidth fibres.

Tim Birks, Jonathan Knight and Philip Russell at the University of Bath in England are thanked for collaborations, enlightening arguments, and their willingness to share their expertise and data.

We thank David Webb, Helen Dobb and Karen Carroll at Aston University in the UK for their help with writing gratings in our fibres.

John Harvey, David Hirst and Laura Harvey at the University of Auckland in New Zealand have assisted us in many ways, particularly in the development of cleaving and measurement techniques.

Lili Wang and her colleagues at the Xi'an Institute of Optics and Precision Mechanics in China are thanked for providing us with preforms made by a variety of techniques.

Many people read and helped with the preparation of this manuscript. We would like to particularly thank the several long-suffering people who innocently expressed an interest in our work, only to be landed with a manuscript of this book for comment. The thoughtful suggestions and careful proof-reading of John Love, Roger Tanner, Rod Vance, Haida Liang, Cristiano Cordeiro and Boris Kuhlmey were very much appreciated. Further thanks are due also due to our colleagues for their patience in reading various drafts, especially Alex Argyros.

We would also like to thank Maksim Skorobogatiy from the École Polytechnique de Montréal for material on the all-polymer Bragg fibres.

Financial support for the research described in this book was provided by a number of sources. We gratefully acknowledge the Australian Research Council, the New Zealand Foundation for Research Science and Technology, the Australian Photonics Cooperative Research Centre, The Australian Academy of Technological Sciences and Engineering and the Bandwidth Foundry Pty Ltd - the latter being a Major National Research Facility supported by the Australian and New South Wales Governments.

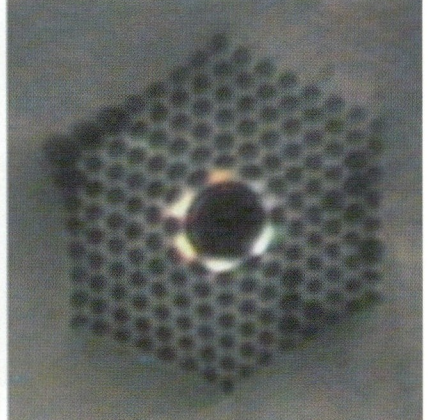

Figure 3.4: A low contrast photonic bandgap fibre shows the interaction of rod and core modes. The wavelengths of light in the core correspond to bandgaps in the cladding structure.

Figure 8.7: The solid region surrounding the core of a hollow core fibre can support modes, as shown here, which can couple light out of the core.

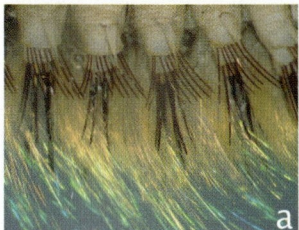

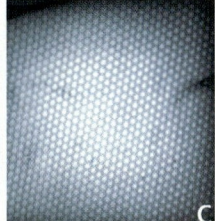

Figure 3.5: Photonic bandgap effects seen in the spines of a seamouse (a) and (b) and a cross-section through a spine showing the microstructure (c).

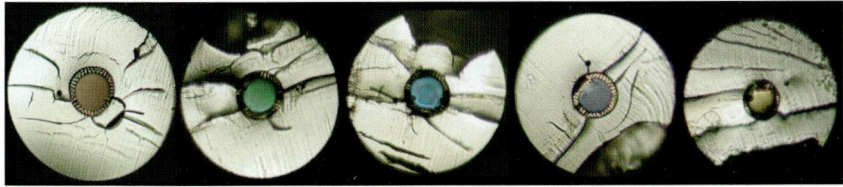

Figure 8.5: Different colours guided in the hollow core of a bandgap fibre drawn to different diameters, changing through red, green, blue, blue-violet and yellow (= red + green) as the structure size decreases.

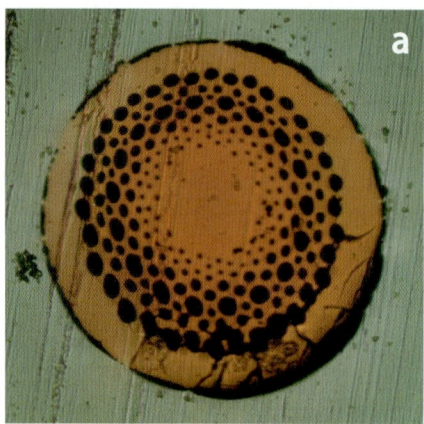

Figure 11.2: (a) Cross-section of a doped mPOF. Orange coloured parts are Rhodamine doped.

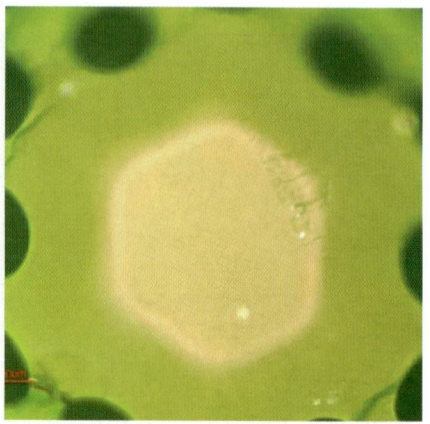

Figure 11.3: Cross-section of a preform prior to the dopant diffusion fronts meeting at the core. The green is the fluorescence of the dye.

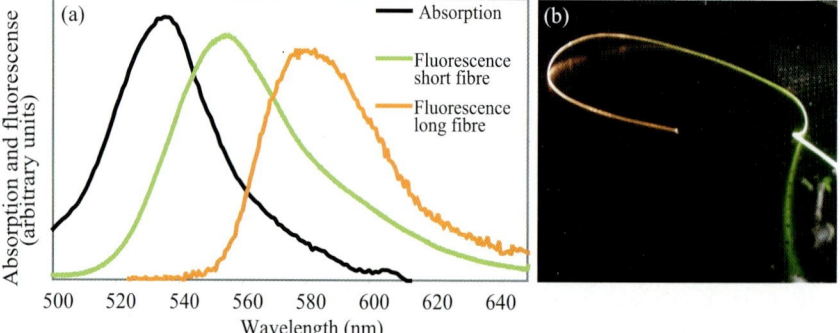

Figure 11.5: (a) The absorption spectrum of the Rhodamine 6G dye and two fluorescence spectra corresponding to short (a few mm) and long (2 m) fibres. (b) Photograph illustrating the re-absorption process.

Figure 11.7: Silica nanoparticles containing R6G and Rhodamine B and their appearance under normal light and UV light.

Contents

List of Figures

List of Tables

List of Symbols and Acronyms

β	propagation constant
ε	aspect ratio
κ	thermal conductivity
Λ	period
μ	viscosity
ρ	spot size
ρ	density
σ	surface tension
χ	relative hole size
Bi	Biot number
Br	Brinkman number
Ca	capillary number
C_o	collapse ratio
Dr	draw ratio
Fr	Froude number
n_g	group index
n_{eff}	effective index
Pe	Peclet number
Re	Reynolds number
T_{g}	glass transition temperature
T_{draw}	drawing temperature
v_{g}	group velocity
v_{ph}	phase velocity
V	V parameter, dimensionless frequency
ARROW	anti-resonant reflective optical waveguide
DMA	differential mode attenuation
DMD	differential mode delay
FBG	fibre Bragg grating
EH	hybrid mode

FTTH	Fibre-to-the-Home
GI POF	graded-index POF
GImPOF	graded-index mPOF
HC-mPOF	hollow-core mPOF
HE	hybrid mode
LAN	local area network
LD	laser diode
LED	light emitting diode
LPG	long period grating
MFD	mode field diameter
MOF	microstructured optical fibre
mPOF	microsturctured polymer optical fibre
NA	numerical aperture
PMMA	polymethylmethacrylate
POF	polymer optical fibre
PCF	photonic crystal fibre
RZ	Return-to-Zero
NRZ	Non-Return-to-Zero
TE	transverse electric mode
TM	transverse magnetic mode
VCSEL	vertical cavity surface emitting laser

History and Applications of Polymer Fibres and Microstructured Fibres

What's past is prologue.

William Shakespeare, *The Tempest*

This chapter places the rest of the book in context. It describes the history and state-of-the-art of both polymer fibres (POFs) and microstructured optical fibres (MOFs). The physical properties of these fibre types differ considerably in terms of the materials used and the possible waveguide geometries, and these form the basis for the difference in their applications. This chapter outlines both the physical differences and the major applications of each. The applications of POFs are described in more detail because most of the applications of microstructured fibres reappear in later chapters.

1.1 Considerations In The Early Stages Of Polymer Fibre Development

Historically the important drawback of polymer fibres has been their relatively low transmission compared to silica. This has, more than any other issue, dominated the development of POF. It has spurred materials development to reduce attenuation, and led to a gradual appreciation of the alternative virtues of POF, such as their ability to remain flexible even with large cores, the variety of fabrication techniques available for such fibres, and the fact that they can incorporate many forms of dopant material.

The struggle to reduce transmission losses has been a dominating theme during the development of all optical fibres. In the very earliest days, while the major loss mechanisms were still unclear, there was a great deal of experimentation with the materials used. In the early 1950s, some of the very first optical fibres were made of polymer, with claddings made with liquid beeswax [Hecht 1999]. The coating was later changed to a liquid polymer which was cured and painted black to prevent light leaking out. The requirement for painting is itself highly suggestive of the poor quality of the optical guidance.

In parallel to this polymer work, optical fibres were also being made of silica, sometimes with polymer claddings. For more than a decade after these early studies, the transmission of optical fibres of all types remained surprisingly bad. When, in 1965 Charles Kao concluded that a loss of 20 dB/km was needed for them to be practical for data transmission, that seemingly modest figure was still 50× lower than what was possible using the best fibres at the time [Hecht 1999]. Fortunately, Kao also supplied the insights into how to achieve these lower losses. By greatly improving the purity of silica, he and others were able to dramatically reduce the loss of silica fibres. The best silica fibres now have a loss of some 0.15 dB/km at 1550 nm and form the backbone of modern telecommunications systems.

This extraordinary success was deeply problematic for polymer optical fibres, because the absorption loss of polymers is intrinsically much higher than that of silica. Losses in all optical fibres are dominated at short wavelengths by Raleigh scattering, but in polymers, absorption due to the harmonics of the C-H vibration becomes very significant at wavelengths longer than about 600 nm. One approach to reducing this has been to shift the harmonics to longer wavelengths by replacing hydrogen with something heavier, such as deuterium or fluorine. The use of fluorination has substantially improved the transmission of polymer fibres, not just by reducing the loss, but also by extending their transmission window into the infrared. This is particularly attractive because it allows the use of low-cost components such as sources and detectors previously developed for use with silica fibres. This advantage is offset by the additional cost of fluorination, always a difficult and expensive process, which produces hydrofluoric acid as a by-product.

Theoretically, the best fluorinated material should have a loss approaching that of silica [Koike 1998], though the best experimental results are about only 10 dB/km. For a variety of reasons, the most commonly used polymer for POF remains polymethylmethacrylate (PMMA) which has a theoretical loss limit of 106 dB/km at the most useful transmission window (650 nm). Other polymers that have been used for POF include Polycarbonate and Polystyrene, the former being used in applications that require higher thermal stability. Figure 1.1 compares the transmission of fibres made from silica, fluorinated polymer (the proprietary material CYTOP), and PMMA. Figure 1.2 shows the transmission spectrum of PMMA in its transparency region.

Despite their high attenuation, POF have continued to be developed commercially because they have some major advantages over silica. Figure 1.3 shows the transmission characteristics of PMMA based POF from the first commercially available fibre, produced by DuPont in 1963. Processing improvements lowered the loss for PMMA step index fibres to around 150 dB/km at 650 nm in the 1980s. Graded-index POF was first made in 1982 which by 1990 had achieved similar transmission characteristics to step index POF.

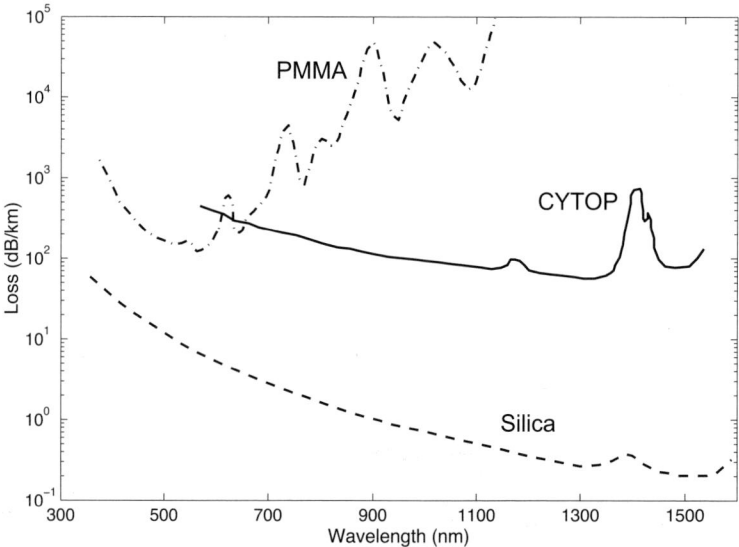

Fig. 1.1. The transmission of fibres made from silica, fluorinated polymer (the proprietary material CYTOP), and PMMA. After Murofushi [1996].

1.2 Fabrication Considerations For Polymer And Glass Optical Fibres

Fabrication methods may not initially seem an important point of comparison between polymer and glasses, but in fact there are significant differences between them which impact strongly on their applications. These are more than simply cost related as the fabrication methods also define what kind of fibres can be made using the two platforms. Understanding these constraints is particularly significant for microstructured fibres, as in some cases these allow the production of fibre types that would be very hard to produce by any other means.

There are two general approaches to making optical fibres. In most cases, particularly in silica, fibres are drawn from a "preform" – a short, fat version of the fibre which contains the desired radial structure (see Chapter 5). In other cases, the fibre is drawn directly from liquid material. In glasses, this liquid is simply molten glass [Palais 1992], but for polymer fibres it may either be molten polymer, or unpolymerised material [Daum et al. 2002].

Glass fibres are normally produced using preforms. The desired refractive index profile is usually produced by doping with small amounts of materials such as germanium or boron. The most important doping methods are based around vapour deposition, in which layers of the desired material are successively deposited and oxidised by a flame. This can be done in a very controlled manner to produce a wide range of index profiles with very high

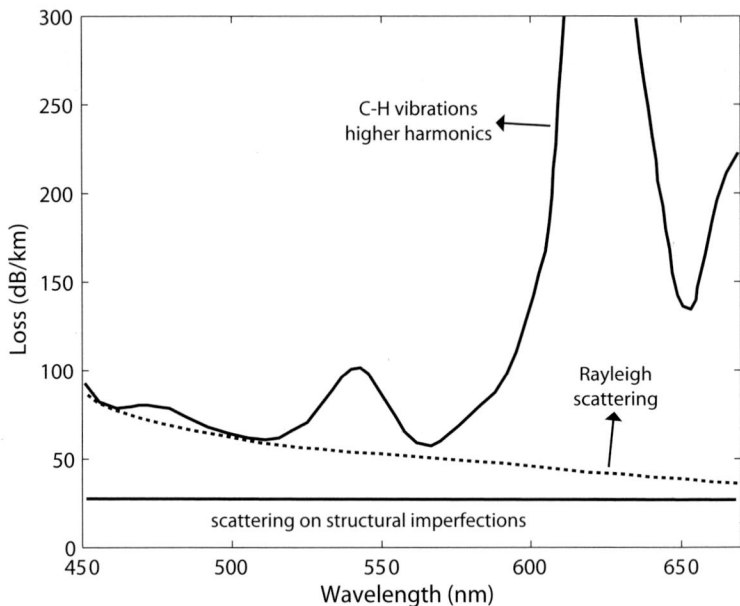

Fig. 1.2. The transmission of a PMMA based POF in its transparency region. After Kaino [1992].

purity materials. Indeed, the invention of modified chemical vapour deposition [MacChesney et al. 1974] revolutionised silica fibre optics and allowed a wide variety of fibres to be reliably produced.

Not all glass fibres however, can be made using vapour deposition techniques because they require the precursor material to be volatile at room temperature. Other approaches to making glass fibres include the "double crucible" technique, in which two molten glasses are drawn together, and the "rod-in-tube technique", in which a tube of one material is collapsed onto a rod of the other material. Although conceptually simple, these techniques typically do not produce good interfaces between the two materials, with stress and scattering at the interface being common problems. It is interesting to note that these very problems spurred the first experiments into microstructured fibres [Kaiser et al. 1973, Kaiser and Astle 1974, Marcatili 1973], before the development of vapour deposition techniques rendered them unnecessary.

In polymers, by contrast, there is no equivalent to vapour deposition and it is difficult to control the refractive index profile precisely. Some of the approaches that have been tried include co-extrusion of materials with slightly different composition, centrifuging during polymerisation and interfacial gel polymerisation [Daum et al. 2002]. Co-extrusion comes closest to vapour deposition techniques in terms of its versatility, as it allows multiple steps in refractive index to be achieved. In all cases however, considerable optimisation of the process is required to obtain the required profile. This is due in no

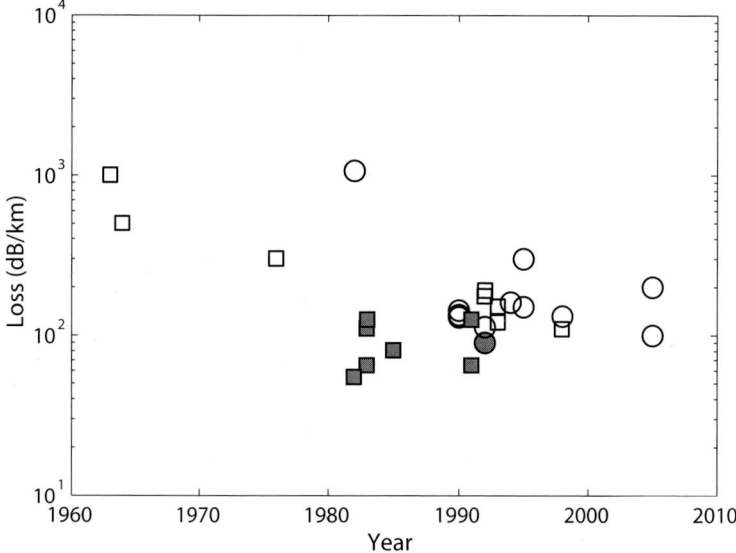

Fig. 1.3. The development of PMMA based POF. Step-index fibres are shown as squares, while graded-index fibres are shown as circles. Two transparency regions have been used, around 570 nm (shaded) and 650 nm (hollow).

small part to diffusion which occurs in polymers at a far lower temperature than in glasses, and can cause the refractive index profile to change. The glass transition temperature (the point at which the main chain of the polymer becomes mobile) in polymers is typically in the range 100 - 150 °C, while the softening temperature for silica is over 1500 °C.

In practice most polymer fibres are step index in nature, often made using different but compatible polymers. These can include copolymers (polymers made from two or more different monomers) or completely different polymers, such as polycarbonate and amorphous Teflon. Step-index fibres can be produced by a wide variety of methods, such as extrusion, casting one material around the other and coating and curing the cladding from a liquid. The use of two different materials generally produces much larger differences in refractive index than doping.

Thus glasses and polymers have almost complementary qualities. For glass, at least in silica, there are excellent methods for producing doped fibres with small index contrasts between core and cladding, while polymer fibres can be easily mass produced, but generally as high-contrast step-index fibres. More complex refractive index profiles can be produced but the generic techniques require careful optimisation.

1.3 Index Contrast Between Core And Cladding

The contrast between the core and cladding of an optical fibre has considerable impact on its optical properties. The most obvious impact is on the light collecting ability which is largely determined by its numerical aperture (NA):

$$NA = \sqrt{n_{\mathrm{co}}^2 - n_{\mathrm{cl}}^2}, \tag{1.1}$$

where n_{co} and n_{cl} are the core and cladding refractive indices, respectively. Polymer optical fibres, which usually have a high contrast between core and cladding, have relatively high NAs. Their usual range is from 0.2 - 0.5, with NAs of up to 0.7 being commercially available [Daum et al. 2002].

In silica, single-mode fibres completely dominate applications from data transmission to sensing. This has not been the case for polymer, where it has been difficult to make single-mode POF, for exactly the same reason that it is easy to make them with a high numerical aperture.

The number of modes an optical fibre supports is related to the V parameter (the normalized frequency) given by:

$$V = \left(\frac{2\pi a}{\lambda}\right) NA \tag{1.2}$$

where a is the core radius and λ is the free space wavelength. If V is less than 2.405 the fibre is single-moded [Palais 1992]. Thus, there is an implicit trade-off between the refractive index contrast (controlling the NA) and core size. A small refractive index contrast, which could be produced by doping, allows a larger core but dopant diffusion must be controlled if the refractive index profile is to remain stable. Fibres with larger refractive index contrasts are more robust in this respect but require a much smaller core, which increases the scattering loss at the core/cladding interface. The situation for POF is further complicated if the fibres are required to be single-moded in the transparency region of the polymer, as this implies an even smaller core than for the infrared wavelengths used in silica.

Doped PMMA based single-mode polymer optical fibres have been made with a loss of approximately 200 dB/km at 652 nm [Koike and Nihei 1991]. Since this initial work however, no other group seems to have been able to emulate this result. Single-mode POF has recently become commercially available but with very high loss. The most common application of single mode POF is for writing gratings, which is usually done in the infrared, where phase masks are easily available. Typically, 10 cm of a single mode PMMA fibre has a loss of 15 dB at 1500 nm (including splice loss) [Liu et al. 2006]. Most are not single-moded in the visible.

1.4 Physical Properties Of Polymers

Some of the merits of polymer are implicit in their material properties. Representative values of some of these are given for PMMA and silica in Table 1.1. These show that PMMA is much lighter and more flexible than silica, and much more sensitive to temperature changes (a vice or a virtue depending on the application). These trends apply to the majority of polymers, although fluorinated polymers differ in having much higher densities.

Some important features of POFs relate to cost and handling. The fact that they remain highly flexible even with large cores makes them very suitable for applications where ease of connection is an issue. Fibres with small cores, such as single-mode fibres, require submicron precision in alignment while fibres with large cores (many polymer fibres have cores almost 1 mm in diameter) are much more tolerant. In many applications too, it is important that POF are cheap enough to be disposable, do not produce shards when they break, and can operate in the visible. These qualities make them suitable for domestic and/or medical use, and by people with little experience in using optical fibres.

Table 1.1. A comparison of material properties for PMMA and silica.

Property	PMMA	Silica	Units
Density	1195	2200	kg/m^3
Young's Modulus	3.2	72	GPa
Elastic Limit	10%	1%	
Thermal Expansion Coefficient	9×10^{-5}	5.5×10^{-7}	K^{-1}
Thermo-Optic Coefficient	-1.10×10^{-4}	9.2×10^{-6}	K^{-1}

Finally, POFs are compatible with a range of materials that are likely to become increasingly important. Organic materials cannot be incorporated into glass fibres during processing because the high temperatures used in fibre drawing causes them to denature. A large variety of specialised polymeric materials has been developed for applications including nonlinear optics, photorefractives, lasers and emission devices [Kajzar and Swalen 1996, Agronovitch et al. 1996, Charra et al. 2003]. These are much more easily integrated with POF than other fibres, not simply because they could potentially be incorporated into the fibre but because their whole technology platform is based on a similar approach, with an emphasis on flexibility, low processing costs, and using similar materials. Another major growth area for POF is likely to be in biomedical applications, as biological materials too are much more compatible with organic polymers than with glasses. The surfaces of polymers are readily functionalised with materials of biological interest.

1.5 Current Applications Of Conventional Polymer Fibres

1.5.1 Illumination

Illumination optics is the single largest application of POF, and certainly one that exploits its advantages. POF illumination fibres have very large cores, a high light gathering capacity and are highly flexible. Historically most illumination applications have used POF as point sources, for example as highly localised spotlights in buildings or areas where there is limited space. A growing market however is being explored for sidescattering POF for strip lighting. These are produced by the incorporation of scattering particles in the polymer, and can be used at lengths of up to 5 m, usually in conjunction with a LED source.

1.5.2 Sensing

A huge variety of sensors have been made using POF, including those for strain and mechanical deformation [Kuang et al. 2002, Kuang and Cantwell 2003], temperature [Liu et al. 2003], humidity [Muto et al. 1994] and the presence of ozone [O'Keeffe et al. 2005] amongst others. POF sensors are increasingly moving from laboratory prototypes to commercial devices, with the automotive industry currently being the major customer. They are used, for example, for pedestrian protection systems, and seat-occupancy recognition [Polishuk 2006].

Reviews of POF sensors [Bartlett et al. 2000, Peng 2002, Peng et al. 2005] highlight the role of fibre modifications in enabling or improving their performance. Important modifications include writing Fibre Bragg Gratings (FBG) or Long Period Gratings (LPG), doping the core or cladding with fluorescent dyes and producing fibre tapers to increase the evanescent field.

1.5.3 High Speed Data Transmission

There is no question that the most significant growth area for POF is short distance, high speed data transmission. There is increasing demand for more bandwidth in applications as diverse as the automotive industry, consumer electronics as well as Fibre To The Home (FTTH), and Local Area Networks (LAN). In these applications, the ease with which fibre can be connected and installed is very important, and represents a much more critical cost requirement than that of the fibre itself.

Until quite recently step-index POF could satisfy most data transmission applications. They provide data transmission up to 400 Mbits/s over 50 m, and are used in consumer electronics and industrial electronics, as well as the automotive industry. The latter has been described as a "killer application" for

POF, with the market expected to grow to US\$ 567 million by 2008 [Polishuk 2006].

As bit rates increase however, step-index fibres become inadequate, and graded-index (GI) fibres become the obvious replacement (the mechanism by which a GI-fibre improves the bandwidth is discussed in detail in Chapter 9). The most important applications for GI POF derive from a simple fact: while many people have optical fibre in their street, few have it in their homes or offices. This relatively short connection therefore has the capacity to enable a huge increase in usable bandwidth.

The demand for bandwidth is growing, in some cases with strong governmental encouragement. It is anticipated that telephone, internet and television (video on demand as well as broadcasts) will be carried over the same connection within the next decade (the so-called "Triple Play"). This will require data transmission rates of 10 Gbits/s over 100 m. The development of High Definition Television (HDTV) and large video screens represents a further opportunity, requiring data rates of 3 - 5 Gbits/s over 30 m for connecting displays with players. Internet server "farms" also have enormous data transmission requirements, and updating the current copper-based systems with optical fibres would lead to substantial cost savings.

Japan currently leads the world in the installation of FTTH networks, with more than 4 million households connected by 2005, with the whole country expected to be connected by 2008 - 2010. The Japanese Government has also announced that it will accept digital broadcasting television over FTTH from 2006. Korea Telecom aims to complete 10 million FTTH connections by 2010. In Europe, a new consortium (POF-ALL : Paving the Optical Future with Affordable Lightning Fast Links) is developing POF for FTTH applications.

While the development of FTTH seems inevitable, the best system used to implement it is less clear. A number of different approaches are currently being explored. In some cases silica fibre is used as backbone within the building (such as an apartment block) with wireless or copper cables providing the connection to the consumer. Fluorinated high bandwidth POF operating in the infrared has also been used. The POF-ALL consortium by contrast, aims to use PMMA based POF operating in the visible, with links of 100 Mbit/s transmission over 300 m at 520 nm, and 1 Gbit/s over 100 m at 650 nm. [Nocivelli 2006].

Apart from the obvious optical requirements of bandwidth and transmission over sufficient length, considerations that will influence the final choice of fibre system include the durability of the fibre, its ability to withstand tight bends, and installation cost, including ease of connection and cheap sources and detectors. Many people in the field believe that large-core fibres operating in the visible are required for "domestic" applications.

Clearly, the uptake of FTTH and other short distance, high bandwidth applications present enormous opportunities for POF, but they also impose much higher technical requirements than were previously needed. In an effort to define these requirements and bring greater consistency to the industry, an additional set of international standards (IEC SC86A) was established, which took effect in the latter part of 2006. These are given in Table 1.2, and better than anything else, articulate the direction POF is taking in terms of high speed data transmission. They define new standards (here denoted 1 - 4). The first is anticipated to be a PMMA-based fibre, while the remaining three would require the use of a fluorinated polymer.

Table 1.2. New IEC SC86A standard for graded-index POF.

Application	1 Digital A/V Data	2 Industrial Mobile	3 Digital A/V LAN	4 Multi-Gbps Data
Loss (dB/km)				
@ 650 nm	< 180	< 100	< 100	n/a
@ 850/1300 nm	n/a	< 40	< 33	< 33
Core Diameter (μm)	500 ± 30	200 ± 10	120 ± 10	62.5 ± 5
Outer Diameter (μm)	750 ± 45	490 ± 10	490 ± 10	250 ± 5
Bandwidth (Gbps/100m)				
@ 650 nm	> 0.4	> 1.6	> 1.6	n/a
@ 850/1300 nm	n/a	3–8	3.76–10	3.76–10

1.6 History And Overview Of Microstructured Fibres

Microstructured optical fibres (MOFs), also known as "photonic crystal fibres" (PCF) or "holey" fibres [Russell 2006], produce their light guidance effects through a patterning of tiny holes which run along the entire length of the fibre. If the features of the microstructure are made sufficiently small, then the microstructured region may be considered homogeneous, and the inclusion of air or other materials allows the average refractive index to be varied quite simply. A cladding region that includes tiny air holes will have a depressed refractive index compared to the solid core, allowing the conditions for total internal reflection, or, equivalently, mode confinement, to be easily satisfied.

This approach to controlling the refractive index was explored many years ago as a way to avoid chemical doping [Kaiser et al. 1973, Kaiser and Astle 1974, Marcatili 1973], and to allow sensing of low refractive index materials [Vali and Chang 1992]. Neither of these initial forays into the use of microstructures in fibres seems to have been developed further, and indeed they suggested that MOFs are not very different from conventional fibres, albeit with a much larger refractive index contrast.

The much more interesting reality was not understood until many years later, when Philip Russell and his colleagues at the University of Bath in the

UK happened on the idea again, this time spurred by the growing interesting in planar photonic crystals. The fibre that more than any other spurred the development of the field was their endlessly single-mode fibre [Knight et al. 1996, Birks et al. 1997].

Fig. 1.4. The original endlessly single-mode fibre [Knight et al. 1996]. *Photo courtesy of the University of Bath, UK.*

A suite of new discoveries and applications followed swiftly. By making the arrangement of holes, or the holes themselves, have different profiles in the x and y directions highly birefringent (HiBi) or polarisation maintaining fibres were fabricated [see e.g. [Ortigosa-Blanch et al. 2000, Issa et al. 2004]). Not only can very high birefringence be obtained by this means, but the temperature variation of the birefringence is far less in MOFs than it is in conventional HiBi fibres [Issa et al. 2004] because they do not rely on stress.

The dispersion properties of MOFs have also attracted attention (see e.g. [Knight et al. 2000, Reeves et al. 2002, Saitoh et al. 2003]), particularly because of the possibilities they offer for dispersion compensation. By changing the core size of the fibre, it is also possible to make them have either very low or very high optical nonlinearity [Broderick et al. 1999]. An appealing use of this property was the development of a fibre system for supercontinuum generation [Ranka et al. 2000], which has many of the attractive features of a laser (such as high intensity and coherence) but over a wide spectral range.

By making the bridges between the holes extremely thin, it is possible to make fibres which are essentially "air clad", and such fibres can have numerical apertures greater than 0.9 [Wadsworth et al. 2004].

A particularly significant feature of MOFs is that they offer the ability to guide light in air or other low index materials through the photonic bandgap effect [Cregan et al. 1999]. Photonic bandgap effects are most simply understood in one-dimensional structures such as multilayer stacks which can be designed to reflect particular wavelengths. These wavelengths are described as lying within the "bandgap" of the structure: they cannot be transmitted, and so are reflected. If we imagine this multilayer stack to be rolled up into a cylinder, we obtain a "Bragg fibre", in which the wavelengths within the bandgap are reflected by the multilayer cladding and are transmitted along the hollow core.

Exactly this approach has been used to produce "Swiss roll" Omniguide fibres, in which a two material multilayer is rolled up to produce a hollow core fibre [Fink et al. 1998, Kuriki et al. 2004], see Fig. 1.6(a). Obtaining a very high refractive index contrast requires the use of very different materials. This raises issues of material compatibility, which are particularly significant during the draw process. The materials used in these fibres include a combination of polymers such as poly(ether imide) and poly(ether sulfone) with soft glasses such as arsenic triselenide. The refractive indices of these materials range from 1.62 for poly(ether sulfone) to 2.82 for arsenic triselenide. More exotic structures have also been produced in the fibre geometry which include metals and semi-conductors [Bayindir et al. 2004].

All-polymer Bragg fibres have also been explored, with both hollow and solid cores in the visible [Skorobogatiy 2005b], mid-IR [Dellemann et al. 2003] and THz regimes [Skorobogatiy and Dupuis 2007]. Unlike the Omniguide fibres, these fibres have a relatively low refractive index contrast. Typically the fibres combine PMMA (with a refractive of 1.49) with polystyrene (with a refractive index of 1.56). Applications include: ultra-sensitive and compact surface plasmon resonance sensors [Hassani and Skorobogatiy 2006, Skorobogatiy and Kabashin 2006]; ultra-high bandwidth multimode fibres and components for short-range datacom applications [Skorobogatiy and Guo 2007, Skorobogatiy 2005a]; and biocompatible or biodegradable fibres for in-vivo light delivery and sensing [Dupuis et al. 2007]. Bandgap guidance at different wavelengths in the visible has been observed in solid-core Bragg fibres with different diameters drawn from the same preform (see Fig. 1.5). Losses of these fibres vary between 5-20 dB/m. Main contribution to such loss is believed to be scattering from imperfections in the fibre. There is considerable scope for improving this loss with refinements to the fabrication process.

Finally, an approach which requires only a single material is to make a two-dimensional microstructure which approximates a Bragg fibre. The structure consists of rings of holes [Argyros et al. 2001, Argyros 2002, Argyros et al.

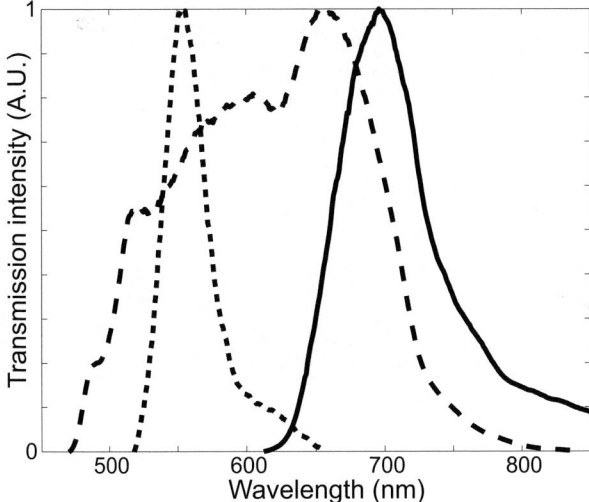

Fig. 1.5. The transmission of three different all-polymer Bragg fibres showing the presence of band gaps. *Figure courtesy of Maksim Skorobogatiy, École Polytechnique de Montréal.*

2004, Vienne et al. 2004], see Fig. 1.6(b), or more commonly a two-dimensional array of holes (as shown in Fig. 1.6(c) and 1.7) is used to produce the bandgap. As the light is predominantly in the hollow core and not the solid material, this allows guidance of wavelengths that would be absorbed by the material. It also allows guidance in materials of low refractive index, such as liquids or gases, as it does not rely on total internal reflection.

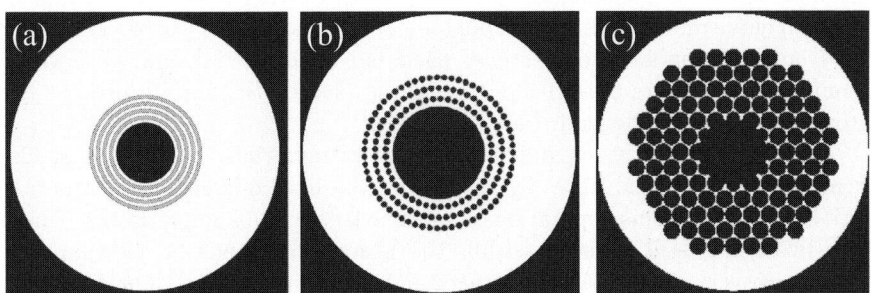

Fig. 1.6. Schematic diagrams of the cross-section of a hollow-core (a) Bragg fibre (b) ring-structured Bragg fibre and (c) a photonic bangap fibre. Air regions are shown in black.

These diverse properties are not only valuable in isolation, they can also be combined in new ways, to make for example, large core fibres with very

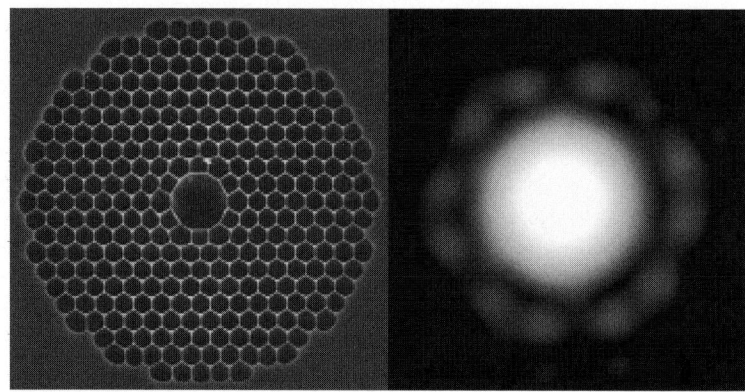

Fig. 1.7. (a) Electron micrograph of an air guiding photonic bandgap fibre and (b) optical near-field image of guiding in the hollow core. *Image courtesy of Crystal Fibre A/S, Blokken 84, 3460 Birkerod, Denmark http://www.crystal-fibre.com*

high numerical apertures for high power laser applications [Wadsworth et al. 2003, Limpert et al. 2003], or fibres in which the holes not only produce light guidance but can also introduce other materials: gases or liquids for sensing or low threshold Raman processes [Benabid et al. 2002b, 2005, Fini 2004, Ritari et al. 2004], or to produce tuneable effects [Mach et al. 2002]. MOFs have also improved evanescent field sensing. They can be designed to have high intensity evanescent fields (that is, a relatively large fraction of the light is in the evanescent field), and the incorporation of the material to be tested into the holes of the microstructure allows physically robust sensors to be made, while keeping the material in close proximity to the core. This application is being developed for a range of biological sensing applications [Myaing et al. 2003, Jensen et al. 2005, Emiliyanov et al. 2007]. Other novel applications of MOFs include multicore fibres for imaging and interconnects [van Eijkelenborg 2004] and the guidance of particles [Benabid et al. 2002a], and potentially atoms, in hollow core fibres using the light guided by the fibre to prevent the atoms from hitting the wall [Dall et al. 2002].

MOFs have greatly expanded the range of materials that can be realistically used for optical fibres by separating the light guidance from the material properties. This separation is most extreme in bandgap fibres, where transmission is still possible within the absorption bands of the material, but is also very significant in averaged-index guiding fibres. Many fibre systems suffer from issues associated with the interface between different materials [Monro et al. 2000], or the presence of dopants. These can adversely affect the mechanical properties of the fibre, as they do in soft glasses, and they can also make the resulting fibres vulnerable to degradation. In polymer fibres, dopant diffusion is one of the factors limiting the temperature stability. Indeed, temperature stability may make MOFs a superior solution even in applications where conventional fibres work well, such as birefringent fibres. While the loss of MOFs remain above those of conventional fibres, their performance is improving rapidly. For solid core fibres the best loss figure now stands at 0.28 dB/km [Zhou et al. 2005], and for hollow core fibres at 1.2 dB/km [Roberts et al. 2005]

Although most work in microstructured fibres to date has been in silica, other materials are increasingly being explored, including a range of soft glasses [Kiang et al. 2002, Kumar et al. 2002, 2003, Feng et al. 2005]. These new materials will expand further the rich field of MOF applications by allowing more highly nonlinear materials to be used, and by using materials that guide in difficult wavelength ranges such as the mid-infrared. Single-mode microstructured polymer optical fibres (mPOF) were first made some years ago [van Eijkelenborg et al. 2001] and have become one of the most interesting offshoots of the field. MPOF have now been made using a number of different materials, including Topas [Emiliyanov et al. 2007]and biodegradable materials [Dupuis et al. 2007].

The diversity of fabrication processes available for polymers makes them easier to use for complex structures and to scale up for commercial production. As well as the mechanical advantages discussed earlier, polymers also offer considerable scope for development through doping, grafting or inclusion of nanoparticles. They open up a range of applications that silica fibres would probably never address, such as large core fibres for high bandwidth transmission (Fig. 1.8) or applications that require the fibre to be cheap and disposable. Some of the more important applications of mPOF are explored in the latter part of this book.

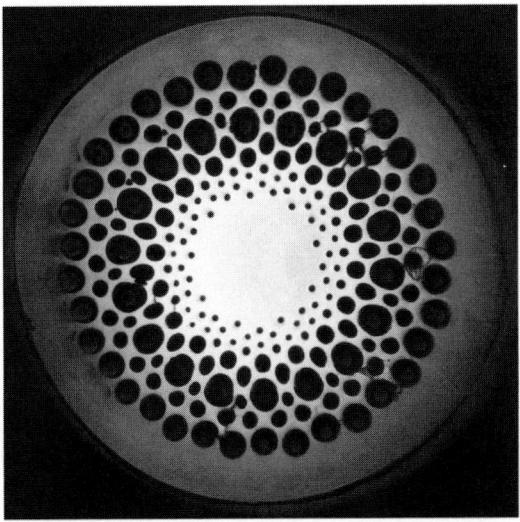

Fig. 1.8. A graded-index mPOF. The outer diameter of the fibre is 250 microns. After Large et al. [2006].

References

Agronovitch, V M, Kajzar, F, and Lee, C Y-C (1996). *Photoactive Organic Materials: Science and Applications.* Kluwer Academic Publishers, Dordrecht, The Netherlands.

Argyros, A (2002). Guided modes and loss in Bragg fibre. *Optics Express*, 10(24):1411–7.

Argyros, A, Bassett, I M, van Eijkelenborg, M A, and Large, M C J (2004). Analysis of ring-structured Bragg fibres for TE mode guidance. *Optics Express*, 12(12):2688–98.

Argyros, A, Bassett, I M, van Eijkelenborg, M A, Large, M C J, Zagari, J, Nicorovici, N A P, McPhedran, R C, and de Sterke, C M (2001). Ring structures in microstructured polymer optical fibres. *Optics Express*, 9(13):813–20.

Bartlett, R J, Philip-Chandy, R, Eldridge, P, Merchant, D F, Morgan, R, and Scully, P J (2000). Plastic optical fibre sensors and devices. *Transactions of the Institute of Measurement and Control*, 22(5):431–57.

Bayindir, M, Sorin, F, Abouraddy, A F, Viens, J, Hart, S D, Joannopoulos, J D, and Fink, Y (2004). Metal-insulator-semiconductor optoelectronic fibres. *Nature*, 431.

Benabid, F, Couny, F, Knight, J C, Birks, T A, and Russell, P St J (2005). Compact, stable and efficient all-fibre gas cells using hollow-core photonic crystal fibres. *Nature*, 434(7032):488–91.

Benabid, F, Knight, J, and Russell, P St J (2002a). Particle levitation and guidance in hollow-core photonic crystal fiber. *Optics Express*, 10(21):1195–1203.

Benabid, F, Knight, J C, Antonopoulos, G, and Russell, P St J (2002b). Stimulated Raman scattering in hydrogen-filled hollow-core photonic crystal fiber. *Science*, 298(5592):399–402.

Birks, T A, Knight, J C, and Russell, P St J (1997). Endlessly single-mode photonic crystal fiber. *Optics Letters*, 22(13):961–3.

Broderick, N G R, Monro, T M, Bennett, P J, and Richardson, D J (1999). Nonlinearity in holey optical fibers: measurement and future opportunities. *Optics Letters*, 24(20):1395–7.

Charra, F, Agronovitch, V M, and Kajzar, F, editors (2003). *Organic Nanophotonics.* Springer.

Cregan, R F, Mangan, B J, Knight, J C, Birks, T A, Russell, P St J, Roberts, P J, and Allan, D C (1999). Single-mode photonic band gap guidance of light in air. *Science*, 285:1537–9.

Dall, R G, Hoogerland, M D, Tierney, D, Baldwin, K G H, and Buckman, S J (2002). Single mode hollow optical fibres for atom guiding. *Applied Physics B: Lasers and Optics*, 74(1):11–8.

Daum, W, Krauser, J, Zamzow, P E, and Ziemann, O (2002). *POF Polymer Optical Fibers for Data Communication.* Springer Verlag, Berlin, Germany, first edition.

Dellemann, G, Engeness, T D, Skorobogatiy, M, and Kolodny, Uri (2003). Perfect mirrors extend hollow-core fiber applications. *Photonics Spectra*, 37:60.

Dupuis, A, Guo, N, Gao, Y, Godbout, N, Lacroix, S, Dubois, C., and Skorobogatiy, M. (2007). Porous double-core biodegradable polymer optical fiber. *Optics Letters*, 32:109.

Emiliyanov, G, Jensen, J B, Bang, O, Hoiby, P E, Pedersen, L H, Kjær, E M, and Lindvold, L (2007). Localized biosensing with Topas microstructured polymer optical fiber. *Optics Letters*, 32(5):460–462. Erratum: p. 1059.

Feng, X, Mairaj, A K, Hewak, D W, and Monro, T M (2005). Nonsilica glasses for holey fibers. *Journal Lightwave Technology*, 23(6):2046–54.

Fini, J M (2004). Microstructure fibres for optical sensing in gases and liquids. *Measurement Science and Technology*, 5:1120–8.

Fink, Y, Winn, J N, Fan, S H, Chen, C P, Michel, J, Joannopoulos, J D, and Thomas, E L (1998). A dielectric omnidirectional reflector. *Science*, 282(5394):1679–1682.

Hassani, A and Skorobogatiy, M (2006). Design of the microstructured optical fiber-based surface plasmon resonance sensors with enhanced microfluidics. *Optics Express*, 14:11616.

Hecht, J (1999). *City of Light: The Story of Fiber Optics*. Oxford University Press, UK.

Issa, N A, van Eijkelenborg, M A, Fellew, M, Cox, F, Henry, G, and Large, M C J (2004). Fabrication and study of microstructured optical fibers with elliptical holes. *Optics Letters*, 29(12):1336–8.

Jensen, J, Hoiby, J P, Emiliyanov, G, Bang, O, Pedersen, L, and Bjarklev, A (2005). Selective detection of antibodies in microstructured polymer optical fibers. *Optics Express*, 13(15):5883–9.

Kaino, T (1992). Chapter 1. In Hornak, L A, editor, *Polymers for lightwave and integrated optics*. Marcel Dekker, New York.

Kaiser, V P and Astle, H W (1974). Low-loss single-material fibers made from pure fused silica. *Bell System Technical Journal*, 53:1021–39.

Kaiser, V P, Marcatili, E A, and Miller, S E (1973). A new optical fiber. *Bell System Technical Journal*, 52(2):265–9.

Kajzar, F and Swalen, J D, editors (1996). *Organic Thin Films for waveguiding Nonlinear Optics*. Taylor & Francis, Gordon and Breach. Advances in non-linear optics Vol 3.

Kiang, K M, Frampton, K, Monro, T M, Moore, R, Tucknott, J, Hewak, D W, Richardson, D J, and Rutt, H N (2002). Extruded single mode non-silica glass holey optical fibres. *Electronics Letters*, 38(12):546–7.

Knight, J C, Arriaga, J, Birks, T A, Wadsworth, W J, and Russell, P St J (2000). Anomalous dispersion in photonic crystal fiber. *IEEE Photonics Technology Letters*, 12(7):807–9.

Knight, J C, Birks, T A, Russell, P St J, and Atkin, D M (1996). All-silica single mode optical fiber with photonic crystal cladding. *Optics Letters*, 21(19):1547–9.

Koike, Y (1998). POF from the past to the future. In *Proceedings of the International Plastic Optical Fibres conference*, volume 7, pages 1–8, Berlin, Germany.

Koike, Y and Nihei, E (1991). Low loss graded index and single mode polymer optical fiber. In *ACS Polymer Preprints - Photonic Polymer for Device Applications*, volume 32, pages 111–112, New York, USA.

Kuang, K S C and Cantwell, W J (2003). The use of plastic optical fibre sensors for monitoring the dynamic response of fibre composite beams. *Measurement Science and Technology*, 14:736–45.

Kuang, K S C, Cantwell, W J, and Scully, P J (2002). An evaluation of a novel plastic optical fiber sensor for axial strain and bend measurements. *Measurement Science and Technology*, 13:1523–34.

Kumar, V V Ravi Kanth, George, A K, Knight, J C, and Russell, P St J (2003). Tellurite photonic crystal fiber. *Optics Express*, 11(20):2641–5.

Kumar, V V Ravi Kanth, George, A K, Reeves, W H, Knight, J C, Russell, P St J, Omenetto, F G, and Taylor, A J (2002). Extruded soft glass photonic crystal fiber for ultrabroad supercontinuum generation. *Optics Express*, 10(25):1520–5.

Kuriki, K, Shapira, O, Hart, S, Benoit, G, Kuriki, Y, Viens, J, Bayindir, M, Joannopoulos, J, and Fink, Y (2004). Hollow multilayer photonic bandgap fibers for nir applications. *Optics Express*, 12(8):1510–7.

Large, M C J, Ponrathnam, S, Argyros, A, Bassett, I, Punjari, N S, Cox, F, Barton, G W, and van Eijkelenborg, M A (2006). Microstructured polymer optical fibres: New opportunities and challenges. In Burillo, G, Ogawa, T, Rau, I, and Kajzar, F, editors, *Molecular Crystals and Liquid Crystals Journal, Special issue, Proceedings of the 8th international conference on frontiers of polymers and advanced materials*, volume 446, pages 219–31. Taylor & Francis.

Limpert, J, Schreiber, T, Nolte, S, Zellmer, H, Tünnermann, A, Iliew, R, Lederer, F, Broeng, J, Vienne, G, Petersson, A, and Jakobsen, C (2003). High-power air-clad large-mode-area photonic crystal fiber laser. *Optics Express*, 11(7):818–23.

Liu, H Y, Liu, H B, and Peng, G D (2006). Polymer optical fibre Bragg gratings based fibre laser. *Optics Communications*, 266(1):132–5.

Liu, H Y, Liu, H B, Peng, G D, and Chu, P L (2003). Observation of type I and type II gratings behavior in polymer optical fiber. *Optics Communications*, 220(4-6):337–43.

MacChesney, J B, O'Connor, P B, and Presby, H M (1974). A new technique for preparation of low-loss and graded index optical fibers. *Proceedings of the IEEE*, 62(9):1280–1.

Mach, P, Dolinski, M, Baldwin, K W, Rogers, J A, Kerbage, C, Windeler, R S, and Eggleton, B J (2002). Tunable microfluidic optical fiber. *Applied Physics Letters*, 80(23):4294–6.

Marcatili, E A J (1973). Air clad optical fiber waveguide. US Patent 3712705.

Monro, T M, West, Y D, Hewak, D W, Broderick, N G R, and Richardson, D J (2000). Chalcogenide holey fibres. *Electronics Letters*, 36(24):1998–2000.

Murofushi, M (1996). Low loss perfluorinated POF. In *Proceedings of the International Plastic Optical Fibres conference*, pages 17–23, Paris, France.

Muto, S, Sato, H, and Hosaka, T (1994). Optical humidity sensor using fluorescent plastic fiber and its application to breathing condition monitor. *Japanese Journal of Applied Physics*, 33(10):6060–4.

Myaing, M T, Ye, J Y, Norris, T B, Thomas, T, Jr, J R Baker, Wadsworth, W J, Bouwmans, G, Knight, J C, and Russell, P St J (2003). Enhanced two-photon biosensing with double-clad photonic crystal fiber. *Optics Letters*, 28(14):1224–6.

Nocivelli, A (2006). Plastic fibre promises ubiquitous optical access. *FibreSystems Europe in association with LIGHTWAVE Europe*, page 14.

O'Keeffe, S, Fitzpatrick, C, and Lewis, E (2005). Ozone measurement in visible region: an optical fibre sensor system. *Electronics Letters*, 41(24):1317–9.

Ortigosa-Blanch, A, Knight, J C, Wadsworth, W J, Arriaga, J, Mangan, B J, Birks, T A, and Russell, P St J (2000). Highly birefringent photonic crystal fibers. *Optics Letters*, 25(18):1325–27.

Palais, J C (1992). *Fiber Optic Communications*. Prentice Hall, Englewood Cliffs, New Jersey, USA.

Peng, G D (2002). Prospects of POF and grating for sensing. In *Proceedings of the International Conference on Optical Fiber Sensors*, volume 1, pages 714–6, Portland, USA.

Peng, G D, Liu, H Y, Chu, P L, and Wang, T (2005). Sensor applications of polymer optical Bragg gratings. In *Proceedings of the International Conference on Polymer Optical Fiber*, volume 14, pages 213–6, Hong Kong, China.

Polishuk, P (2006). Plastic optical fibers branch out. *IEEE communications Magazine*.

Ranka, J K, Windeler, R S, and Stentz, A J (2000). Visible continuum generation in air-silica microstructure optical fibers with anomalous dispersion at 800 nm. *Optics Letters*, 25(1):25–7.

Reeves, W, Knight, J C, Russell, P St J, and Roberts, P (2002). Demonstration of ultra-flattened dispersion in photonic crystal fibers. *Optics Express*, 10(14):609–13.

Ritari, T, Tuominen, J, Ludvigsen, H, Petersen, J, Sørensen, T, Hansen, T, and Simonsen, H (2004). Gas sensing using air-guiding photonic bandgap fibers. *Optics Express*, 12(17):4080–7.

Roberts, P J, Couny, F, Sabert, H, Mangan, B J, Williams, D P, Farr, L, Mason, M W, Tomlinson, A, Birks, T A, Knight, J C, and Russell, P St J (2005). Ultimate low loss of hollow-core photonic crystal fibres. *Optics Express*, 13(1):236–44.

Russell, P St-J (2006). Photonic-crystal fibers. *Journal Of Lightwave Technology*, 24(12).

Saitoh, K, Koshiba, M, Hasegawa, T, and Sasaoka, E (2003). Chromatic dispersion control in photonic crystal fibers: application to ultra-flattened dispersion. *Optics Express*, 11(8):843–52.

Skorobogatiy, M (2005a). Design principles of multi-fiber resonant directional couplers with hollow Bragg fibers: example of a 3x3 coupler. *Optics Letters*, 30:2849.

Skorobogatiy, M (2005b). Efficient anti-guiding of TE and TM polarizations in low index core waveguides without the need of omnidirectional reflector. *Optics Letters*, 30:2991.

Skorobogatiy, M and Dupuis, A (2007). Ferroelectric all-polymer hollow Bragg fibers for terahertz guidance. *Applied Physics Letters*, 90:113514.

Skorobogatiy, M and Guo, N (2007). Bandwidth enhancement by differential mode attenuation in multimode photonic crystal Bragg fibers. *Optics Letters*, 32:900.

Skorobogatiy, M and Kabashin, A V (2006). Photon crystal waveguide-based surface plasmon resonance bio-sensor. *Applied Physics Letters*, 89.

Vali, V and Chang, D B (1992). Low index of refraction optical fiber with tubular core and/or cladding. US Patent 5155792.

van Eijkelenborg, M A (2004). Imaging with microstructured polymer fibre. *Opics Express*, 12(2):342–6.

van Eijkelenborg, M A, Large, M C J, Argyros, A, Zagari, J, Manos, S, Issa, N A, Bassett, I, Fleming, S, McPhedran, R C, de Sterke, C M, and Nicorovici, N A P (2001). Microstructured polymer optical fibre. *Optics Express*, 9(7):319–27.

Vienne, G, Xu, Y, Jakobsen, C, Deyerl, H-J, Jensen, J, Sørensen, T, Hansen, T, Huang, Y, Terrel, M, Lee, R, Mortensen, N, Broeng, J, Simonsen, H, Bjarklev, A, and Yariv, A (2004). Ultra-large bandwidth hollow-core guiding in all-silica Bragg fibers with nano-supports. *Optics Express*, 12(15):3500–8.

Wadsworth, W J, Percival, R M, Bouwmans, G, Knight, J C, Birks, T A, Hedley, T D, and Russell, P St J (2004). Very high numerical aperture fibers. *IEEE Photonics Technology Letters*, 16(3):843–5.

Wadsworth, W J, Percival, R M, Bouwmans, G, Knight, J C, and Russell, P St J (2003). High power air-clad photonic crystal fiber laser. *Optics Express*, 11(1):48–53.

Zhou, J, Tajima, K, Nakajima, K, Kurokawa, K, Fukai, C, Matsui, T, and Sankawa, I (2005). Progress on low loss photonic crystal fibers. *Optical Fiber Technology*, 11(2):101–10.

2

Concepts in Waveguide Theory

In theory, there is no difference between theory and practice. But, in practice, there is.

Jan L. A. van de Snepscheut

This chapter assumes the reader is familiar with the basic undergraduate optical concepts of electromagnetic waves, refractive index, phase, polarisation, diffraction and interference. The focus is on the important concepts, and although in this chapter we only look at conventional fibres we examine the concepts with sufficient clarity to ensure that they will be correctly applied to microstructured fibres. We also discuss whether a concept has a primarily historical, pragmatic or theoretical basis. Mathematical formalism is kept to a minimum while retaining as much rigour as necessary, and most of the approximations presented are based on sound theoretical concepts. We begin with how modes are classified and the role of polarisation and birefringence. We then move on to counting modes and the important concept of cutoff. We look at coupling light into fibres and then at dispersion. Finally, we cover the physics behind how modes are confined in conventional fibres to prepare the reader for the next chapter.

2.1 Modes

A mode is an electromagnetic field configuration or pattern which propagates unchanged down the length of a uniform straight fibre. Thus its intensity distribution is unaffected by such propagation. In the presence of loss, the shape of the intensity distribution does not change but the overall magnitude decreases. Each mode has a phase velocity v_{ph} associated with it; the effective index n_{eff} of the mode is given by

$$n_{\mathrm{eff}} = \frac{c}{v_{\mathrm{ph}}} \tag{2.1}$$

where c is the speed of light. The propagation constant β is a commonly used quantity related to the effective index by

$$\beta = \frac{2\pi}{\lambda}n_{\text{eff}} = kn_{\text{eff}} = \frac{\omega}{c}n_{\text{eff}} \qquad (2.2)$$

where λ is the wavelength, k is the wave number and ω is the angular frequency.

Modes can be classified as bound, radiation, evanescent or leaky based on the nature of the propagation constant. Only bound and leaky modes will be of interest here, although all types are briefly discussed for the sake of completeness. Bound and radiation modes both have *real* propagation constants; they can propagate indefinitely without loss. Bound modes are localised (with most of their power close to and within the core of the fibre). They can have simple or complicated structures as in Figs. 2.1(a) and (b) but the power outside any given radius decays exponentially. Radiation modes are delocalised (see Figs. 2.1(c) and (d)): although the intensity profile decays, it does not decay very rapidly and the power outside any given radius is always infinite. On the other hand, leaky modes have *complex* propagation constants; they can propagate over long distances but decay exponentially at a rate determined by the imaginary part of the propagation constant. Only leaky modes with small decay rates are of interest. Their transverse profiles resemble bound or radiation modes near the core but at large distances increase exponentially as in Figs. 2.1(e) and (f). Evanescent modes have *pure imaginary* propagation constants; they do not propagate at all but decay exponentially within the space of a few wavelengths. There are extremely few situations where they are of any importance.

Conventional fibres support, in principle, all of the above mode types; although, in most situations only bound and radiation modes are relevant. Microstructured fibres, on the other hand, mainly support leaky modes.

2.1.1 Degeneracy, Polarisation And Birefringence

Although each mode has a well defined effective index, it may be possible for two or more modes to share the same effective index. Such modes are said to be degenerate. It is important to distinguish two different ways in which modes can be degenerate. If the degeneracy is caused by an underlying symmetry of the waveguide then the modes will be degenerate at all wavelengths. Readers may remember that a plane wave in free space can always be decomposed into exactly two orthogonal polarisation states. Group theoretical arguments [McIsaac 1975a,b] reveal that the same result is true in optical fibres: symmetry-induced degenerate modes can also always be decomposed into exactly two orthogonal polarisation states. For example, standard step-index telecommunications fibres will support a single pair of orthogonally polarised degenerate bound modes if the fibre is perfectly circularly symmetric.

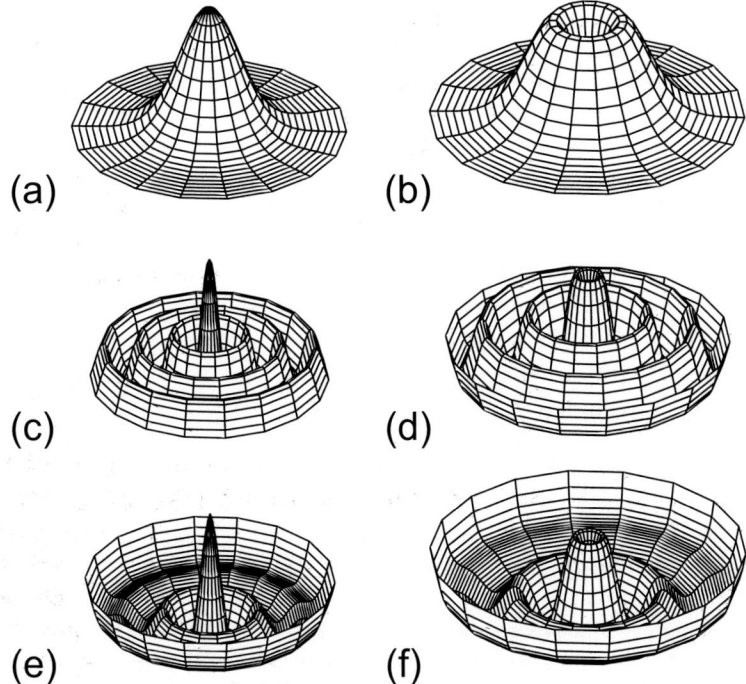

Fig. 2.1. Schematic description of the transverse intensity profile of different classes of modes. Modes (a) and (b) represent a fundamental and higher-order bound mode. Both are localised and decay rapidly away from the axis. Modes (c) and (d) represent two different radiation modes. The fields are delocalised and decay slowly. Modes (e) and (f) are leaky modes. They resemble bound or radiation modes near the axis but their intensity diverges at large distances.

Accidental degeneracy occurs when modes have the same effective index but there is no symmetry responsible. This is likely to only happen at specific wavelengths called crossing points. On either side of the crossing point these modes will be nondegenerate. The intensity profiles and polarisation states of accidentally degenerate modes need not be related in any way.

Degeneracy is very important in understanding the effects of perturbations. Any perturbation which does not preserve a symmetry will break the degeneracy resulting in two modes with close but different effective indices. These two modes will remain similar in almost all their properties (such as intensity profile) but will have extremely different (nearly orthogonal) polarisation states. A fibre is said to be birefringent when modes that have the same (or similar) intensity profiles, but differ in their polarisation properties, travel at different speeds down the fibre. For example, real fibres are never perfectly circularly symmetric. Although the asymmetry may be small, any lack of symmetry induces some polarisation mode dispersion which can be problematic in long-distance applications.

Birefringence can occur through unintended or unavoidable perturbations, but it can also be achieved deliberately to create fibres that control or maintain specific polarisation states. In conventional fibre, it has been straightforward to create useful amounts of birefringence by using material or stress-optic effects to change the polarisation properties of the fibre: such as in hi-bi fibres. In conventional fibres, the use of geometry, such as elliptically shaped cores, has only resulted in small amounts of birefringence. However, geometric effects have been much easier to exploit in microstructured fibres as we see in Chapter 3.

2.1.2 Weak Guidance

Electromagnetism is inherently a vector phenomenon but a good conceptual understanding can often be obtained from various scalar approximations to the full theory. These approximations can explain most properties of modes except those related to polarisation.

Most conventional fibres use a refractive index contrast between core and cladding which is relatively small. For example, in silica step-index fibres the refractive index of core and cladding are about 1.48 and 1.46, respectively. In polymer fibres the contrast is somewhat larger with the values being of order 1.49 and 1.41. In both cases, though, the index difference is small enough to allow the "weak guidance" approximation to be used [Gloge 1971]. This approximation, perhaps more literally a "low contrast" approximation, allows a considerable simplification to be made in the mathematical modelling. In the most general case, solving Maxwell's equations requires that six equations be solved, respectively relating to the three components of E and H. The "weak guidance" approximation reduces this to a single equation: the scalar wave equation [Snyder and Love 1983].

While, many features of MOFs can be deduced from this approach, it is not strictly appropriate because the refractive index contrast between air and silica or polymer is so large that the weak guidance approximation will not yield numerically accurate solutions. The most important property that is affected by this is, of course, the polarisation which is strongly determined by the index contrast at interfaces. Thus, if the polarisation properties of the modes are required, a vector rather than a scalar treatment has to be used. Chapter 4 covers how polarisation and weak guidance affect the calculation and nomenclature of modes.

2.1.3 Counting Modes

One of the most important defining characteristics of a waveguide is the number of modes it supports. Traditionally, the number of modes supported by a fibre means the number of *bound* modes supported by that fibre. For conventional waveguides, the number of bound or guided modes is an unambiguous concept; for microstructured fibres, however, a useful concept of mode count requires a little more insight. A good starting point is the standard V parameter or dimensionless frequency

$$V = 2\pi \frac{a}{\lambda} \sqrt{n_{\mathrm{co}}^2 - n_{\mathrm{cl}}^2}. \tag{2.3}$$

The parameters in this formula are well-defined for a step-index fibre: a is the radius of the core, λ is the wavelength, while n_{co} and n_{cl} are the refractive indices of the core and cladding, respectively. The total number of bound modes is closely related to the V parameter. In particular if $V < 2.405$ the fibre supports only the doubly degenerate fundamental mode and is referred to as a "single-mode" fibre. For larger values of V, the total number of modes is *approximately* given by

$$N_{\mathrm{modes}} \sim \frac{1}{2} V^2. \tag{2.4}$$

This estimate becomes quite accurate for highly multi-mode fibres and provides a useful estimate even for fibres with only a few modes.

The dimensionless V parameter is itself the product of two other dimensionless numbers. The geometric factor $2\pi \frac{a}{\lambda}$ represents the ratio of a typical waveguide dimension to the wavelength: it is related to the transverse spatial area available for modes but also controls the importance of *diffraction*. The numerical aperture $NA = \sqrt{n_{\mathrm{co}}^2 - n_{\mathrm{cl}}^2}$ represents the index contrast in the system: it is related to the range of angular directions available to modes but also controls the importance of *refraction*. The V parameter thus conceptually indicates how waveguiding is achieved by the balance of *diffraction* and *refraction*.

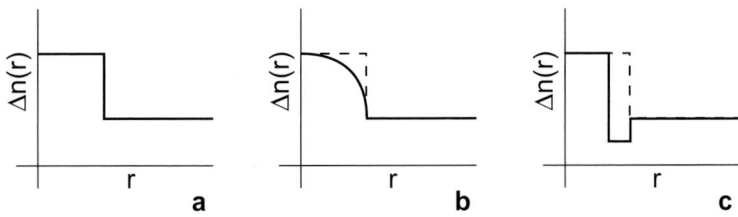

Fig. 2.2. Fibres with various index profiles: (a) Step index (b) Graded index (c) Depressed index. Although all three fibres have the same core radius and core-to-cladding index difference they have different V parameters and thus support different numbers of modes.

We warn the reader that our approach to how to generalise V to non-step index fibres differs from traditional approaches (such as in Snyder and Love [1983]). A productive way to generalise V is revealed by looking at graded index fibres as shown in Fig. 2.2. In quantum mechanics and solid-state physics, the density of states gives the number of distinct modes or states within a small range of energies. In optical waveguide theory, the analogous density of states counts the number of modes within a small range of phase velocities. Although the details are not reproduced here, this can be used [Snyder and Love 1983] to estimate the total number of bound modes and it yields the following integral for a circularly symmetric graded fibre

$$N_{\text{modes}} \sim (\frac{2\pi}{\lambda})^2 \int_0^{\infty} [n^2(r) - n_{\text{cl}}^2] r dr. \tag{2.5}$$

By comparing this more complicated expression to the two previous results for V and N_{modes}, we see that an appropriate generalisation of the V parameter is given by

$$V_{\text{eff}} = \frac{2\pi}{\lambda} \left\{ 2 \int_0^{\infty} [n^2(r) - n_{\text{cl}}^2] r dr \right\}^{1/2}. \tag{2.6}$$

This reduces to the previous result for a step-index fibre. The integral can be evaluated in closed form for any power-law graded-index fibre [Snyder and Love 1983]. For example, when the integral is applied to a parabolic graded-index fibre it predicts exactly half as many modes as a step-index fibre with the same maximum index and radius. Like the simpler expression for V, this estimate is quite accurate for multi-mode fibres yet still provides the correct qualitative insight for single-mode fibres.

Notice that the generalised formula replaces n_{co} with a local index $n(r)$ and then integrates the square of the local numerical aperture over the entire transverse cross-section of the fibre. Unfortunately, this also shows that in general it is difficult to disentangle the angular (numerical aperture) factor from the spatial factor. Nevertheless, this simple estimate allows a deep understanding of both conventional and microstructured fibres.

If we compare all the different profiles shown in Fig. 2.2, we see that as the profile becomes more complicated it is not obvious how to choose an appropriate effective core size or effective index difference. The one unambiguous quantity in all these profiles is the cladding index n_{cl}: the value of the refractive index at large distances from the axis of the fibre. This is important because it is the cladding index that determines which modes are bound (i.e. propagate without loss) and which radiate transversely to infinity.

2.1.4 Effective Indices And Cutoff

Bound modes must have effective indices lying in the range

$$n_{\text{cl}} < n_{\text{eff}} < n_{\text{max}}. \tag{2.7}$$

where n_{max} is the largest index which occurs anywhere in the structure. This allowed range can easily be understood in terms of phase velocities: it implies that the slowest possible phase velocity corresponds to the highest available refractive index, and that the fastest possible phase velocity is the speed beyond which plane waves can radiate into the cladding. Thus the cladding index serves as a cutoff index for bound modes.

Notice that the allowed range of indices for bound modes does not go all the way down to the minimum refractive index. In a depressed-cladding fibre, such as Fig. 2.2(c), there are regions where the local refractive index is below the cladding index $n(r) < n_{cl}$. The existence of such regions may improve the spatial localisation of *some* modes but as Eq. (2.5) reveals, such regions decrease the *total* number of bound modes the fibre supports.

Now, consider how the number of bound modes varies with wavelength. If the refractive indices themselves are not strong functions of wavelength, then as λ increases both V and the number of modes will gradually decrease. Figure 2.3 shows the effective index of each guided mode as a function of wavelength or V parameter. As the wavelength increases, each mode reaches a point where its effective index drops below the cladding index: at this point the mode ceases to be bound and becomes leaky (as indicated by the dashed lines).

However, notice that one mode is never cutoff, and that for V below 2.405 the fibre is single-moded. In conventional fibres the bound mode with the smallest phase velocity (or largest propagation constant) is called the fundamental mode. In most conventional fibres, the fundamental mode is the doubly degenerate HE_{11} mode.

The smallest value of λ at which the fibre only supports the fundamental mode is called the single-mode cutoff wavelength. It is a general result that if $n(r) \geq n_{cl}$ everywhere then the fibre will *always* support at least one mode, and if left unspecified the word cutoff always refers to the wavelength at which the *second* mode ceases to be bound. It is a common mistake to forget that cutoff is a property of the second mode and occurs at the wavelength where the effective index of the second mode satisfies $n_{eff} = n_{cl}$.

However, in depressed-cladding fibres, at sufficiently long wavelengths, even the fundamental mode ceases to be guided. There is no established name for the wavelength where this happens but the phrase fundamental-mode cutoff is sometimes used. In microstructured fibres, the concept of cutoff and the number of modes supported by the fibre require more thought and this idea is taken up again in Section 3.2.1.

2.2 Coupling Light In And Out

Just as important as the number of modes that a fibre supports is the ability to get light into and out of the fibre. This is important for fibre applications where the ability to connect fibres is paramount or where light must be collected from some type of extended source. Not surprisingly, like the density of states, the capture efficiency of a fibre has both spatial and angular components.

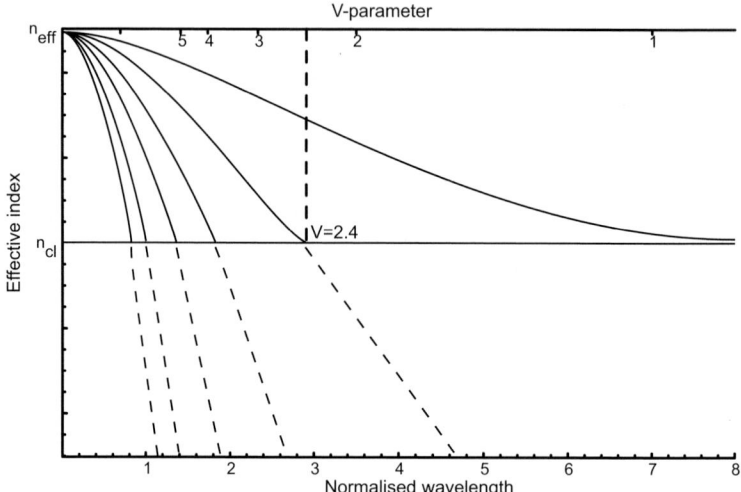

Fig. 2.3. How the number of modes and their effective indices change with either the normalised wavelength or the V parameter and the cutoff wavelengths for a step-index fibre.

2.2.1 Numerical Aperture

When wavelengths are much smaller than the size of spatial structures a geometric optics approach can be used. In the geometric optics limit, a ray will travel down a fibre if its inclination to the core-cladding boundary is less than the critical angle required for total internal reflection. Thus optical fibres will only propagate light that enters within a certain range of angles, known as the acceptance cone (see Fig. 2.4). The vertex half-angle of this cone is called the acceptance angle, $\theta_{\max}$. For strict agreement with other areas of optics, the numerical aperture is *defined* in terms of this angle by

$$NA = n_0 \sin \theta_{\max} \tag{2.8}$$

where n_0 is the refractive index of the surrounding medium (usually air, but sometimes an index-matching liquid) placed at the entrance of the fibre.

For step-index fibres, the above formula yields the same expression we used earlier when counting the number of modes

$$NA = \sqrt{n_{\mathrm{co}}^2 - n_{\mathrm{cl}}^2}. \tag{2.9}$$

However, when this second expression is used as a definition, the relationship between the numerical aperture and the acceptance angle of the fibre becomes only an approximation. This approximation is worst for the case of single-mode fibres.

In multimode fibres, the term *equilibrium numerical aperture* is also used. In this situation $\theta_{\max}$ represents the largest exit angle for all rays assuming an equilibrium mode distribution has been established.(See later in this chapter for a discussion of equilibrium mode distributions.)

This last relation is the basis for an operational definition of numerical aperture through the following characterisation experiment. Light is launched into one end of a fibre and the emerging light is observed at the other end. The light will emerge within a cone. If the fibre is operating within the geometric optics limit, then this cone will have a sharply defined boundary and this experimentally yields the angle θ_{max} needed for the numerical aperture. However, if diffraction effects are large, or the fibre is few-moded, then the boundary of the cone will be diffuse and the numerical aperture will only be approximately determined by this approach.

There are some technical caveats about making sure light is coupled into all the modes and that the fibre is long enough for transient effects to have disappeared (see Section 2.4).

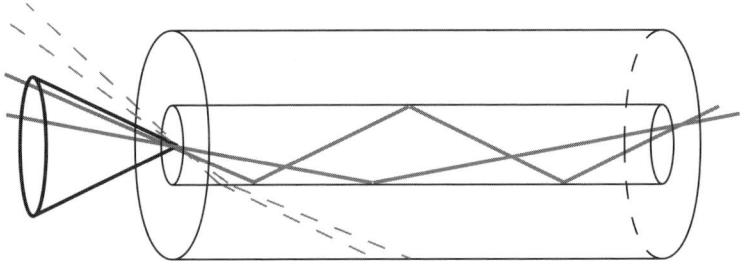

Fig. 2.4. Geometric optics interpretation and origin of numerical aperture. Any rays that enter the fibre within the entrance cone will be guided by total internal reflection within the core, whilst those outside the cone, pass through to the cladding. The half-angle of the cone θ_{max} is related to numerical aperture by Eq. (2.8).

The numerical aperture can of course depend on the wavelength of light and a broad spectrum will usually produce a wider cone with a much less well defined boundary.

2.2.2 Mode Field Diameter Or Spot Size

The transverse cross-sectional area of the fibre core is also crucial in determining how easy it is to launch light into a fibre. The use of the word 'core' already implies that this concept too may be difficult to apply to graded structures. Operationally the amount of loss at a fibre connection for every micron of transverse misalignment is a good measure of how easy a fibre is to connect: however, determining this quantity inevitably requires recourse to numerical calculation or *ad hoc* approximations.

For single or few-moded fibres, the geometric extent of the core is not as important as the *mode field diameter* (MFD) or *spot size* of the modal intensity pattern. There are many definitions of spot size some of which are more useful for characterisation and some for theoretical analysis. A characterisation standard that also has important theoretical meaning is the Petermann II spot size. It is obtained from the second moment (standard deviation) of the far-field pattern of the optical fibre. From diffraction theory, the far-field pattern is the Fourier transform of the near-field pattern (i.e. profile at the output) and so the spot size is inversely related to the width of the far-field pattern. For the mathematically inclined, the Petermann II spotsize ρ can be defined by

$$\frac{1}{\rho^2} = \frac{\int (|\nabla E|^2)dA}{\int |E|^2 dA} \tag{2.10}$$

for weakly guiding waveguides.

The Petermann II spotsize also provides an understanding of the difference between the *effective* index and the naive *average* index defined by

$$n_{average}^2 = \frac{\int n^2 |E|^2 dA}{\int |E|^2 dA}. \tag{2.11}$$

For weakly guiding waveguides one can rigourously show using the variational expression for propagation constant [Snyder and Love 1983] that

$$n_{eff}^2 = n_{ave}^2 - \frac{1}{k^2 \rho^2} = n_{average}^2 - \left(\frac{1}{2\pi}\right)^2 \frac{\lambda^2}{\rho^2}. \tag{2.12}$$

Notice that the effective index is smaller than the naive average index by an amount related to the degree of localisation of the mode.

This formula helps explain one paradox baffling to many: if a mechanism can be found to tightly confine a mode so that ρ is extremely small, it is possible for the effective index (or its real part, at least) to end up less than the minimum index in the fibre. This happens in air core fibres where the effective index can end up being less than one, thus producing *phase* velocities faster than in vacuum.

2.3 Bandwidth And Dispersion

As discussed in Section 1.5.3, one of the major applications for POF is for short-distance, high data rate transmission. Applications such as Fibre-To-The-Home and patch cords for large, high definition television sets require bandwidths of up to several Gigabits/s with a relatively large core to make connectivity easier. In such applications, dispersion can significantly influence pulse propagation. Pulses necessarily contain a range of frequencies, and when these travel at different velocities, the shape of the pulse changes and the relative timing of a stream of pulses is also affected leading to potential detection bit error rates.

In data communication, where pulse trains are initiated at high speed, pulse broadening due to dispersion is a fundamental limit on the maximum transmission rate. In digital systems, the same phenomena that cause pulse spreading also cause variations in the arrival time of pulses, often referred to as *jitter*. Both of these degrade the information content of communication systems.

Bandwidths are usually quoted as a frequency (GHz) or data rate (Gbit/s). In fact the conversion between these quantities depends on assumptions about pulse propagation and type of digital encoding used [Palais 1992], which are often not explicitly stated. Another useful quantity, the bandwidth-length product, is also ambiguous because the way that bandwidth scales with distance depends on the whether the fibre length is above or below the equilibrium length (see Section 2.4).

Data transmission in optical fibres involves encoding a digital (binary) signal as a train of light pulses. Different types of encoding can be used to indicate whether each bit is 1 or 0. The two most common digital encodings are known as Return-to-Zero (RZ) and Non-Return-to-Zero (NRZ). In a RZ encoding, each bit is allocated a time T, with the pulse occupying half of the time slot $T/2$ (the other half is the return to zero). As a result, the maximum possible bandwidth is $\leq 1/T$, and the maximum allowed data rate is $R = 1\text{bit}/T$.

In NRZ encoding the bit occupies the entire pulse length T. In this case the required transmission bandwidth is $\leq 1/(2T)$, half of an RZ system. Thus the maximum allowable data rate using NRZ encoding is $R = 2\text{bit}/T$. This is the most common convention used to convert bit rates to GHz in the literature, and is the one that will be used here.

As the pulses encoding the digital data propagate along the fibre, they are broadened by dispersion. This pulse broadening can be described by a timing delay per unit length $\delta\tau$. For a given frequency modulation $f = 1/T$, the maximum broadening before adjacent pulses start to merge is $\delta\tau \leq T/2$. The factor $\frac{1}{2}$ is a commonly used approximation for a Gaussian pulse, and may vary slightly depending on the impulse response of the fibre [Palais 1992]. We now look at each of the components that contribute to pulse spread $\delta\tau$.

2.3.1 Intra- And Inter-Modal Dispersion

In single-mode fibres, dispersion has two components: that due to the material having a slightly different refractive index at different wavelengths (material dispersion), and that due to the geometrical features of the waveguide (waveguide dispersion). Taken together these effects are termed chromatic or intramodal dispersion, and lead to a pulse spread $\delta\tau_{\text{intra}}$.

In multimode fibres, different modes travel down the fibre at different speeds and so pulses launched into multimode fibres will broaden (or even potentially break up into separate parts). This is called *intermodal* dispersion and leads to a pulse spread $\delta\tau_{\text{inter}}$.

It is usually assumed that intramodal and intermodal dispersion are independent. Thus their effects on pulse spreading combine according to the formula

$$\delta\tau_{\text{total}} = \sqrt{\delta\tau_{\text{intra}}^2 + \delta\tau_{\text{inter}}^2}. \tag{2.13}$$

Each of these contributions is controlled by quite different features of the system. The effect of chromatic dispersion is, of course, proportional to the linewidth of the excitation source used, whereas intermodal dispersion can be strongly influenced by power mixing.

2.3.2 Group Velocity And Chromatic Dispersion

The speed at which a pulse propagates is determined not by its phase velocity but by the group velocity v_{g}. In fibres, the propagation constant will, in general, vary with frequency. The simplest expression for group velocity is given by

$$\frac{1}{v_{\text{g}}} = \frac{d\beta}{d\omega} \tag{2.14}$$

The *group index* n_{g} is a dimensionless quantity analogous to the effective index but it is defined in terms of the group velocity

$$n_{\text{g}} = \frac{c}{v_{\text{g}}}. \tag{2.15}$$

For weakly guiding waveguides and in the absence of material dispersion, the group index is the same as the naïve average index that was discussed earlier in the context of spot-size.

Dispersion in a fibre at a wavelength λ may be characterised by the dispersion coefficient:

$$D_\lambda = -\frac{\lambda}{c} \frac{d^2 n_{\text{eff}}}{d\lambda^2} \tag{2.16}$$

which describes the change in pulse width per nanometer of spectral width per unit distance travelled in the fibre. It is generally quoted in ps/nm/km. Thus, $\delta\tau_{\text{intra}} = D_\lambda \Delta\lambda$ and the total pulse spread can thus be written

$$\delta\tau_{\text{total}} = \sqrt{D_\lambda^2 \Delta\lambda^2 + \delta\tau_{\text{inter}}^2}. \tag{2.17}$$

2.4 Power Mixing In Multimode Fibres

Although most of the important properties of single-mode fibres can be understood by studying ideal fibres, the same is unfortunately not true for multimode fibres. One of the most important ways that a real fibre differs from an idealised one is that the former will suffer from perturbations such as microbending, surface roughness or impurities. Perturbations that are either a result of the fabrication process or arise from external influences will cause an exchange of power between the modes in any multimode fibre. Perturbations can arise from a number of different sources: chemical impurities, small bends in the fibre or roughness at interfaces. Perturbations can be systematically introduced (such as is done to form fibre gratings) or can occur randomly. In this section random perturbations are relevant. These random perturbations cause mode mixing to occur. When light is coupled into a multimode fibre, the launch conditions will determine which modes are initially excited or *filled*. The two most important parameters that determine the initial power distribution are the MFD and *NA* of the source.

2.4.1 Equilibrium

More important than the initial power distribution is the *equilibrium* power distribution. Random perturbations cause power transfer between different modes. After a sufficiently long length of propagation, the relative power distribution among modes becomes statistically constant and continues in this state as it propagates further along the fibre.

The equilibrium distribution depends on the design of the fibre *and* the specific statistical nature of the perturbations. For details on modelling power mixing see Section 4.3. Two properties of the equilibrium distribution are important: it is a steady-state distribution (once achieved or launched it does not change) and it is independent of the initial launch conditions. Power mixing leads to the exponential decay of any departures from equilibrium; thus, the slowest of these exponential decay processes defines the equilibrium length. In practice, the equilibrium length may vary from centimetres to kilometres.

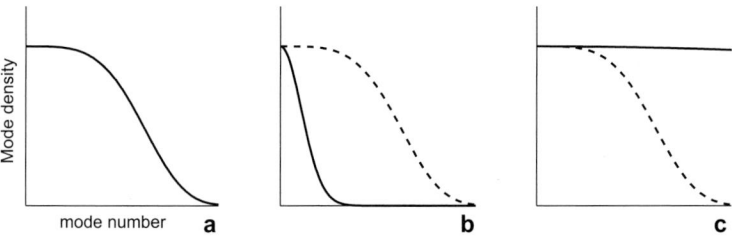

Fig. 2.5. The dashed curves indicate the equilibrium power distribution also shown in (a). The distribution in (b) is under-filled, whereas that in (c) is over-filled.

In equilibrium, each "packet" of energy can be assumed to randomly move between all the different modes spending an amount of time in each that is proportional to the power in that mode. Each packet of energy spends time travelling at both slow and fast group velocities. Thus, on average, all the packets of energy arrive at the same time. This would suggest that once equilibrium is achieved, no further pulse spreading occurs. However, statistical fluctuations about the average will continue to provide a source of pulse spreading. The amount of pulse spreading here will now be proportional to the standard deviation in the group velocities. The standard deviation is also proportional to the square-root of the size of the fluctuations. Thus, in equilibrium, the pulse spreading is proportional to the square-root of the distance travelled along the fibre.

A serious problem that arises in practice is that the perturbations causing mode mixing are of very different magnitudes and an apparent steady state may sometimes be reached via only some of the perturbations [Simard et al. 2003]. Such a steady state distribution will only have one of the two properties of equilibrium mentioned earlier (it will not in general be independent of the launch conditions).

2.4.2 Over- And Under-Filled Launch Conditions

Ideally one should try to launch the exact equilibrium power distribution into a fibre. If the launch power distribution excites the lower-order modes to a greater extent than in equilibrium, then the launching is termed *under-filled*: the power will spread to the higher-order modes until equilibrium is achieved. On the other hand, if the launch power distribution excites the higher-order modes beyond what is necessary for equilibrium, then the launching is termed *over-filled*: the power will spread to the lower-order modes until equilibrium is achieved.

In practice however slightly different definitions are used. The launch is termed under-filled if the source MFD and *NA* are smaller than the fibre core diameter and *NA*; and over-filled if the source MFD and *NA* are larger than the corresponding fibre values.

Thus, as the light travels along the fibre, the energy distribution between the modes evolves until an equilibrium distribution is reached. The length over which this occurs can be measured experimentally (see Section 7.2.6) and is characteristic of the fibre. Silica fibres typically have equilibrium lengths of greater than 1 km, while for polymer fibres they are of the order of a few tens of metres.

Mode-mixing in a multimode fibre has both good and bad aspects. The exchange of energy between fast and slow modes increases the bandwidth by causing the light to travel at an average group velocity. On the other hand, by coupling from low loss modes to ones with higher loss, it may reduce the overall loss of the fibre.

The effect of mode-mixing on pulse spread is shown in Fig. 2.6. Initially, the pulse spread is linear but beyond equilibrium, it has a $\sqrt{L}$ length dependence:

$$\delta\tau_{\text{inter}} = \delta_0 L, \qquad L \ll L_{\text{e}} \qquad (2.18)$$

$$\delta\tau_{\text{inter}} = \delta_0 \sqrt{LL_{\text{e}}}, \qquad L \gg L_{\text{e}} \qquad (2.19)$$

Thus the equilibrium length is an important parameter in determining the way that bandwidth scales with distance.

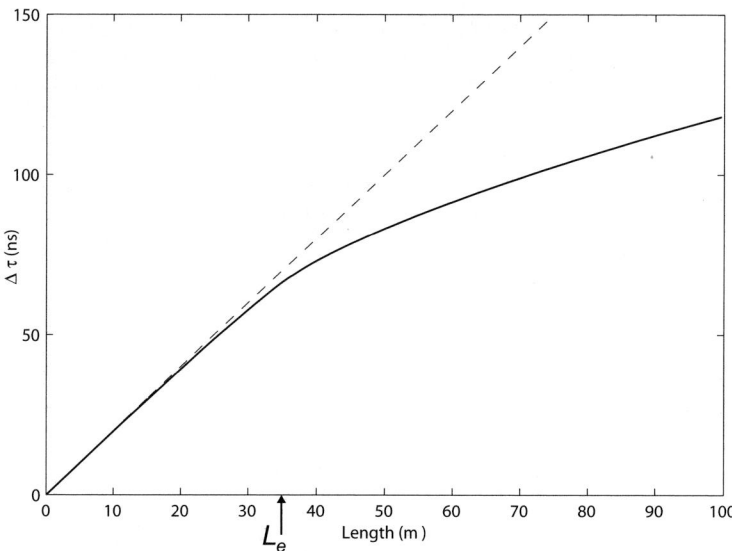

Fig. 2.6. Pulse spread as a function of length in a multi-mode fibre due to intermodal dispersion. The dependence is initially linear before the equilibrium length and has a square-root dependence beyond equilibrium.

2.5 Conventional Guidance Mechanisms

So far in this chapter we have given a descriptive analysis of modes and their properties. Before moving on to microstructured fibres it will be extremely useful to look at the physical mechanisms responsible for guidance in conventional fibres.

Optical fibres guide light using a transverse variation in refractive index. In conventional fibres, the higher refractive index material forms the core of the fibre, as shown in Fig. 2.7(a), allowing light to be guided by total internal reflection. Light incident at the high/low refractive index boundary at an angle greater than the critical angle (as defined by Snell's law) will be totally reflected, giving rise to bound modes. At angles below the critical angle, some of the light is refracted into the cladding, and some is reflected giving rise to cladding modes and leaky core modes, respectively.

Total internal reflection is the basis of almost all optical fibres, including the vast majority of MOFs, the only difference in the latter case being that the refractive index of the cladding is produced using a microstructure rather than by varying the chemical composition of the material forming the fibre.

This is however not the only possible arrangement of two materials of different refractive index, nor is it the only way of making a transparent interface reflective. Just as it is possible to "skim" stones along the surface of water, so it is possible for light at glancing incidence to be guided for a short distance in a structure that has a "core" of a low index material [Fig. 2.7(b)]. This

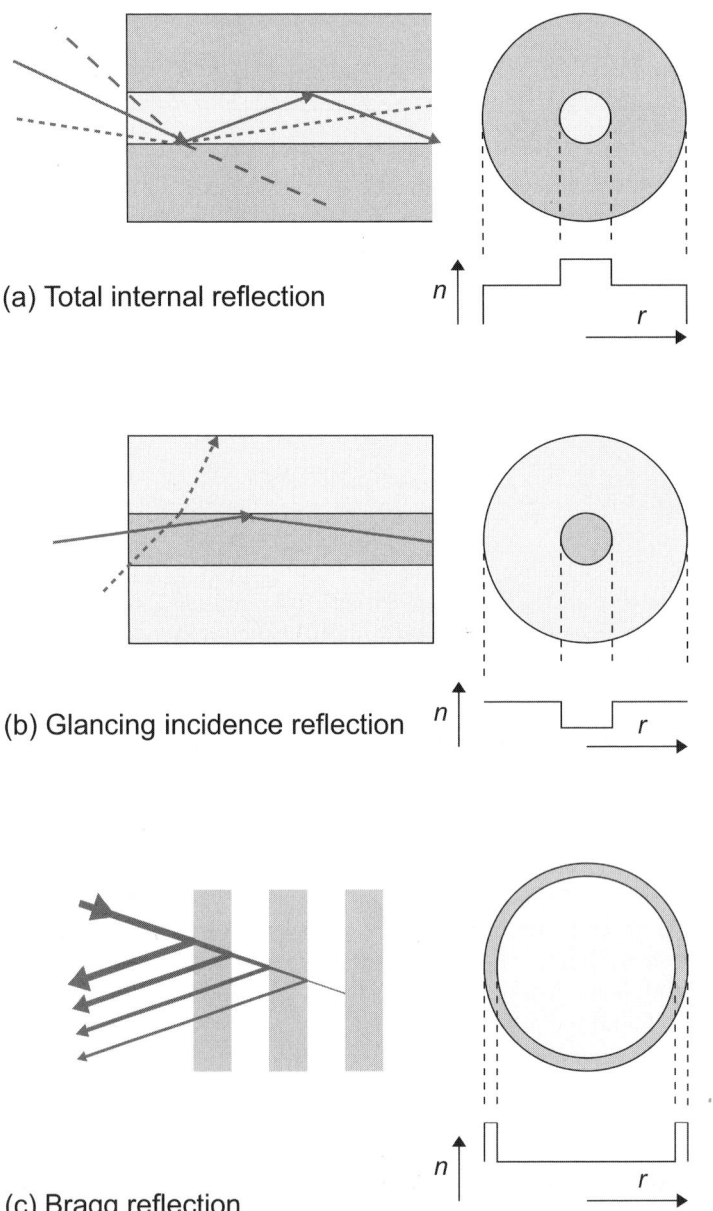

(a) Total internal reflection $\quad n$

(b) Glancing incidence reflection $\quad n$

(c) Bragg reflection $\quad n$

Fig. 2.7. Light guidance by transparent materials. Conventional optical fibres have a raised index core, and guide by total internal reflection (a). When the cladding has a higher index than the core, glancing incidence reflectance is possible (b). Thin film inference effects mean that a thin-walled capillary will give limited reflectance of particular wavelengths (the ARROW effect). If several such capillaries are arranged periodically, these reflectances will add coherently, giving guidance through the Bragg effect (c).

effect does not require any surrounding microstructure and is simply due to the laws of reflection: at large angles of incidence the Fresnel reflection coefficient is close to unity. The modes in such a structure however are extremely leaky, and do not have the discrete spectral features associated with bandgap guidance [Issa et al. 2003]. Much better transmission is obtained through a very thin-walled capillary [Fig. 2.7(c)]. Not only does this structure have twice the number of reflecting interfaces, but at particular wavelengths the light reflected from these interfaces will interfere constructively. Thus, transmission along even this simple structure has discrete spectral features. Although the guidance of a thin-walled capillary is still weak, it exhibits some of the characteristic features associated with bandgap guidance: transmission in discrete wavelength bands which scale with changes in the size of the structure. This guidance mechanism has rarely been used in optical fibres but it does have a long history in planar waveguides and optics more generally. Guiding structures that use it are usually termed **ARROW** waveguides (for Anti-Resonant Reflective Optical Waveguides).

The modes of an unstructured hollow "core" are very leaky because the "cladding" region supports a continuum of modes. Light in the core with any wave vector can therefore couple to a cladding mode and be lost. The effect of introducing a microstructure into the cladding is to break the continuum of cladding modes into a series of discrete bands. The regions between these bands contain no modes and are, by definition, the so called bandgaps. When light in the core has a wave vector that corresponds to a bandgap in the cladding, it will not be able to couple out and thus will remain guided in the core. Outside a bandgap, the wave vector of the core mode will correspond to that of a cladding mode, and coupling to that mode will cause the light to leak out of the core.

The guidance of light in a thin-walled capillary has most of the defining features of bandgap guidance for good reason: the physics behind both effects is the same. The conspicuous differences between capillary guidance and a bandgap fibre are the respective sizes of the cladding region and the transmission loss. There is of course a causal relationship between these features. A concentric arrangement of several capillaries reflects far more of the light incident upon it than does a single capillary and therefore confines light much better in the core.

This section on conventional guidance mechanisms provides the basic physics needed to understand microstructured fibres which are explored in more detail in the next chapter.

References

Gloge, D (1971). Weakly guiding fibers. *Applied Optics*, 10(10):2252.

Issa, N A, Argyros, A, van Eijkelenborg, M A, and Zagari, J (2003). Identifying hollow waveguide guidance in air-cored microstructured optical fibres. *Optics Express*, 11(9):996–1001.

McIsaac, P R (1975a). Symmetry-induced modal characteristics of uniform waveguides - I: Summary of results. *IEEE Transactions Microwave Theory and Techniques*, MTT-23(5):421–29.

McIsaac, P R (1975b). Symmetry-induced modal characteristics of uniform waveguides - II: Theory. *IEEE Transactions Microwave Theory and Techniques*, MTT-23(5):429–33.

Palais, J C (1992). *Fiber Optic Communications*. Prentice Hall, Englewood Cliffs, New Jersey, USA.

Simard, M, Carlson, M, Babin, F, and Tremblay, M (2003). Understanding launch conditions for multimode connector and cable-assembly testing. *EXFO Application Notes*, 092.

Snyder, A W and Love, J D (1983). *Optical waveguide theory*. Chapman and Hall, New York.

3

Guiding Concepts in Microstructured Fibres

"Do not worry if you think it is dark," he said to me, "because I am going to light the light and mangle it for diversion and also for scientific truth."
"Did you say you were going to mangle the light?"
"Wait till you see now."

Flann O'Brien, The Third Policeman.

This chapter builds on the concepts in Chapter 2 but focusses on differences between microstructured and conventional fibres. It begins with a description of the two different guidance mechanisms specific to MOFs. It then goes on to discuss two important ways in which the optical properties of MOFs differ from conventional fibres, namely confinement loss and dispersion. An understanding of these differences is crucial to appreciating how MOFs have expanded the ways and the range in which optical fibres can be used.

The confinement loss and dispersion properties of endlessly single-mode MOFs are considered in some detail. Single-mode fibres have historically been difficult to make in polymers, the example provided gives a good indication how the concepts of confinement loss and dispersion apply to a real fibre.

3.1 Photonic Bandgap Guidance

In a planar geometry, a periodic arrangement of high and low index layers (a multilayer stack) reflects a discrete set of wavelengths. The proportion of light reflected by the structure increases with the number of layers, the refractive index contrast and the uniformity of the stack. None of these properties change when the multilayer stack is rolled into a cylindrical geometry to form the cladding region of a fibre as shown in Fig. 3.1. The cladding reflects exactly the same range of wavelengths as in the planar case, allowing these wavelengths to be transmitted in the hollow core [Argyros 2002]. Increasing the

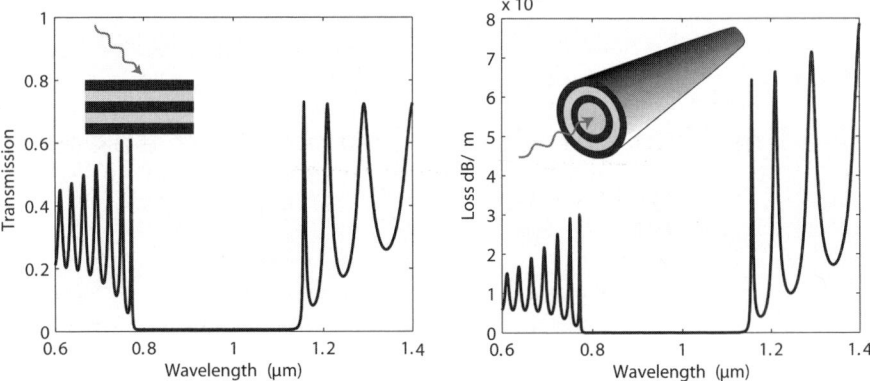

Fig. 3.1. The transmission of a multilayer stack [left] depends on the periodicity of the stack and its refractive index contrast. Some wavelengths are strongly reflected and have almost no transmission through the stack. If the same stack is rolled into a tube [right], the wavelengths that are strongly reflected will be confined to the core and will propagate along its length. The diagram shows representative results, based on a 16 layer stack.

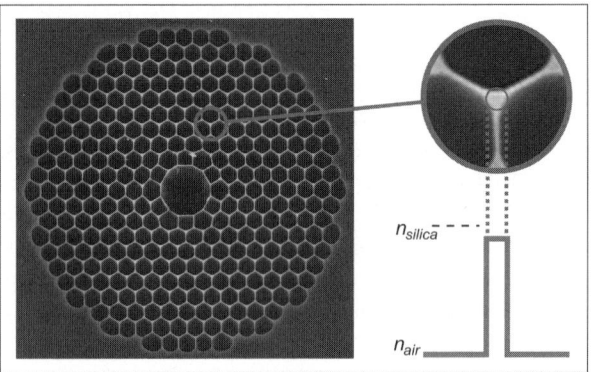

Fig. 3.2. The cladding structure of a photonic bandgap fibre can be thought of as a two-dimensional array of tiny rods suspended in air, each one of which can support modes. *Image courtesy of Crystal Fibre A/S, Blokken 84, 3460 Birkerod, Denmark http://www.crystal-fibre.com*

number of layers or the refractive index contrast, improves the reflectance, and hence the transmission in the fibre. Outside this bandgap region however, light in the core can leak out by coupling to the modes of the cladding structure. An animation illustrating this for a ring structured Bragg fibre is available [Argyros et al. 2004].

The Bragg fibre shown in Fig. 3.1 can also be approximated by using rings of holes [Argyros et al. 2004, 2006, Vienne et al. 2004]. These structures, illustrated together with the more conventional designs, are shown in Fig. 1.6.

The more common two-dimensional photonic bandgap structures (Figs. 1.7) are more challenging to understand intuitively, but exactly the same principles apply: the microstructure in the cladding breaks the cladding modes into a series of discrete bands. In the regions between these bands, light in the core cannot couple to a cladding mode, and must be reflected so that it remains guided in the core.

The origin of cladding modes in these two-dimensional structures can be understood by considering the fibre shown in Fig. 3.2. The solid interstitial regions between the holes (the rods) can be considered as individual cores surrounded by an air "cladding". These tiny rods will have modes just like an optical fibre, and will become single-moded when $V < 2.405$. If we were to plot the effective index as a function of wavelength for the modes of an individual rod we would obtain a dispersion diagram much like Fig. 2.3. If we have an array of similar rods, the dispersion curve for each mode is broadened out into a dispersion band. The width of the band is determined by the degree of interaction or coupling between the rods: at short wavelengths the mode fields in each rod are highly confined and do not couple to other rods; at long wavelengths the rods have a greater interaction with each other. This broadening of the dispersion curves into bands produces a band diagram much like that in the top half of Fig. 3.2). The degree of coupling between rods is also affected by the presence of bridges between the rods: thicker bridges between the rods increase coupling, and hence broaden the bands.

A mode localised in the hollow core at the centre of the structure would have an effective index close to unity. This is shown by the horizontal line in Fig. 3.3. Where this horizontal line crosses the dispersion bands of the rod array the core mode is degenerate or resonant with some mode in the array of rods; power can then couple from the core to the cladding. These high loss regions are shown in the lower half of Fig. 3.3. When the core mode and rod modes have very different effective indices the core mode can propagate with low loss.

As in a Bragg fibre, light is confined in the core when it is unable to couple to any mode of the cladding. This occurs when the effective index of the core mode is very different from that of any cladding mode, that is, when light in the core has a wave vector that corresponds to a bandgap in the cladding. As shown in Fig. 3.3, the edges of the bandgaps in the fibre are approximately at the cutoff frequencies of the rod modes because the index of the cladding for the rods is almost the same as that of the fibre core.

Most photonic bandgap fibres are made using a very high index contrast (e.g. glass or polymer and air), but it is also possible to make such fibres with a very low index contrast [Argyros et al. 2005a,b], for example, by embedding an array of doped silica rods in a pure silica background. Such structures can readily be made by the capillary stacking process (see Section 5.1.2). Indeed,

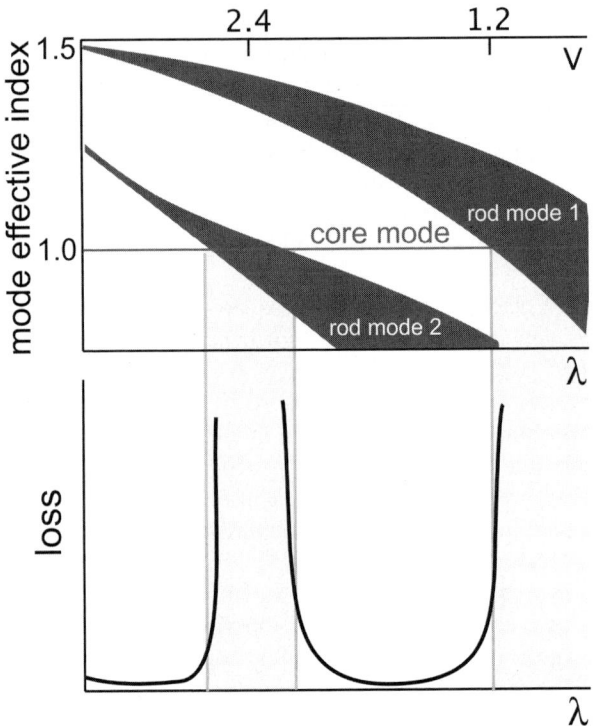

Fig. 3.3. The dispersion bands of the array of rods are shown by the dark shaded regions in the upper diagram. They correspond to the lowest two guided modes of the individual rods. The core is air corresponding to an effective index of 1 and modes guided in the core will have an effective index close to the horizontal line. Coupling between core modes and array modes can explain the regions of high and low loss shown in the lower diagram.

fibres produced by this process are in some ways more "ideal" structures because the high index rods are not joined by bridges. The V parameter can be used to highlight another advantage of reducing the contrast in bandgap fibres. A reduction in index contrast between the core and cladding can be compensated by an increase in structure size while maintaining the same value of V. This also makes it possible to clearly visualise the rod modes and core modes (Fig. 3.4).

Remarkably, structures that are very similar to photonic bandgap fibres have also evolved in nature. A small marine animal known as a seamouse, shown in Fig. 3.5 has highly reflective spines which are due to a hexagonal arrangement of hollow cylinders, exactly equivalent to those found in the cladding of a bandgap fibre. In the structure shown [Parker et al. 2001], 88 layers of cylinders were found (far more than have ever been used in a manufactured bandgap fibre) giving essentially perfect reflectance. This structure is of

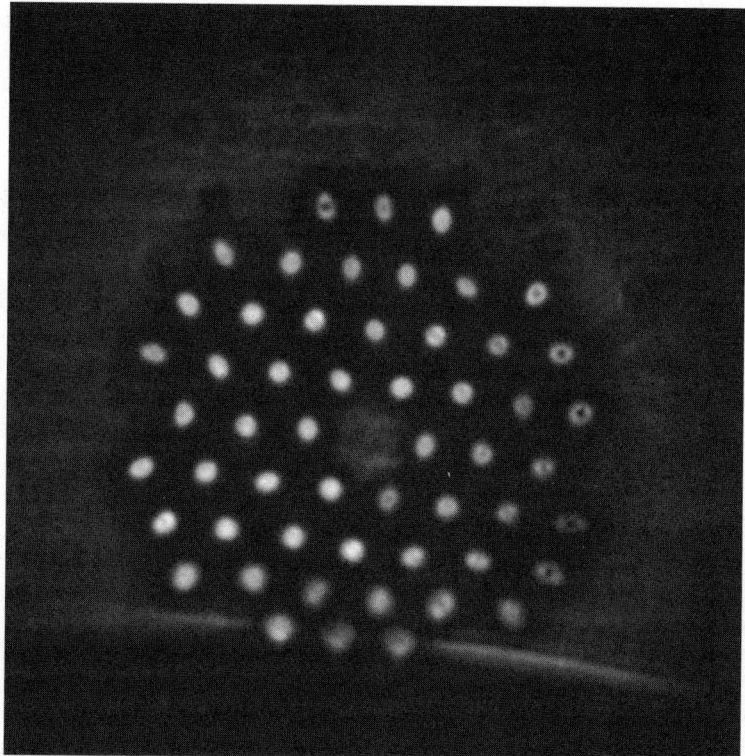

Fig. 3.4. A low contrast photonic bandgap fibre produced using pure and doped silica illustrates the interaction of rod and core modes. The rods guide a different range of wavelengths to the core mode. The wavelengths of light in the core correspond to bandgaps in the cladding structure. The edges of the bandgap correspond to the cutoff frequencies of the rod modes. *Image courtesy of the University of Bath.* See also colour plate at front of book.

course not an optical fibre, as it does not have a "defect" within the array that could act as a core to guide the light. It does however, show quite beautifully the effect of changing the characteristic size of the array. As the structure is tapered, the wavelengths reflected become shorter.

Several points should be added to the discussion at this point, concerning the nature of bandgaps from both a fundamental and a pragmatic perspective. There is a widespread belief that bandgap guidance in fibres requires highly regular periodic structures. This is only partially true.

A rigourous theoretical analysis [Abeeluck et al. 2002, Litchinitser et al. 2002] of Bragg fibres has shown that at short wavelengths the guidance mechanism is through the ARROW mechanism, and is largely independent of the spacing of the cylinders. Periodicity becomes important only at longer wavelengths.

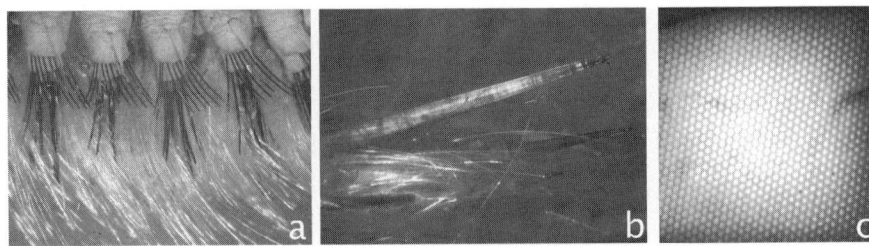

Fig. 3.5. Photonic bandgap effects seen in the spines of a seamouse (a) and (b) and a cross-section through a spine showing the structure responsible for producing the bandgap (c). As the characteristic size of the structure becomes smaller, the wavelength reflected becomes shorter. Thus, the colour changes along the taper of the spine. *Images (a) and (b) courtesy of Greg Rouse of The University of Adelaide.* Image (c) after Parker et al. [2001]. See also colour plate at front of book.

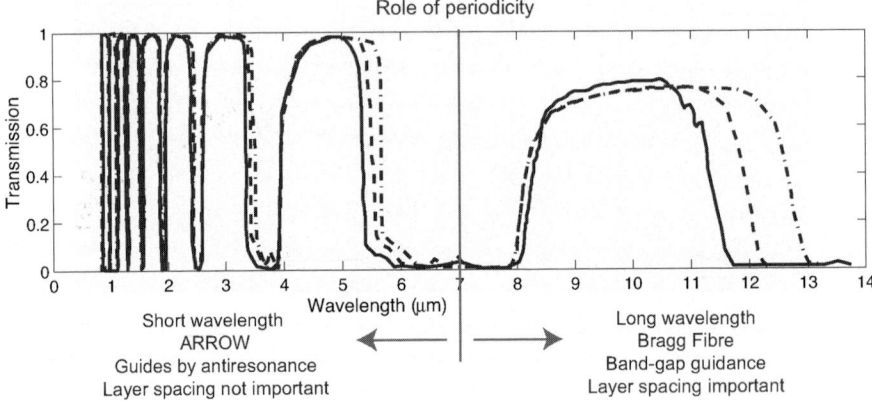

Fig. 3.6. Reflectance mechanisms in Bragg fibres. At short wavelengths only the thickness of the cylinders is critical and reflectance is due to ARROW effects. At longer wavelengths the periodicity of the cylinder spacing becomes important and reflectance is truly due to Bragg effects. After Litchinitser et al. [2002].

Structures in bandgap fibres actually have two types of regularities, the geometry of the high index inclusions (the thickness of the cylinders in the case of the Bragg fibre, or the radius and shape of the rods in the case of the two-dimensional structures), and the low index regions between them. The regularity of the high index regions is significantly more important than the regularity of their spacing, particularly for low index contrasts. The "modes" of the low index regions which separate the inclusions (cylinders or holes) are typically too low in effective index to allow the core mode to couple to them. Thus bandgap fibres operate at least partially via ARROW effects, which do not require a coherent addition of the resonances from the high index regions. True "Bragg" guidance, which does imply a coherent addition of the resonances and hence strict periodicity, becomes significant only at high index contrast or long wavelengths. Indeed, in many cases the performance of bandgap fibres is limited far more by the interface around the central hole than by the structure of the cladding itself (see Chapter 8).

Formally, complete bandgaps, corresponding to a confinement loss of zero or a reflection of 100% of the light, can only occur for infinite cladding structures. As no fibre actually has an infinite cladding, and because limited guidance is possible even in unstructured hollow core fibres or capillaries, there has been some discussion about how to define bandgap guidance, particularly for transmission over short distances. It is clear that bandgap guidance should have discrete spectral features which scale with refractive index contrast and structure size, but exactly the same features and scaling are also associated with the limited guidance possible with a capillary. In the final analysis, the definition of bandgap guidance seems to hinge on loss. In particular, how the confinement loss scales with the number of rings (noting that it should decrease exponentially), or compares to other loss mechanisms (such as material absorption) or to that of a single capillary. In this sense the definition of bandgap guidance seems oddly reminiscent of Louis Armstrong's definition of jazz: "If you have to ask what [it] is, you'll never know".

3.2 Confinement Loss

In MOFs, the matrix material of the core and the cladding is the same, so modes cannot be confined in the sense that they are in conventional (typically doped) fibres. Light is able to leak out or "escape" from the core both through the air holes and through the bridges between them. Loss through the air holes occurs because the holes are relatively small, allowing an evanescent field to tunnel through them [see Fig. 3.7(b)]. This process is equivalent to quantum mechanical tunnelling, and also occurs in conventional fibres with a high index coating when the cladding is sufficiently thin. This kind of confinement loss applies to all modes, and its effect decreases exponentially with the number of rings of holes employed.

Light can also be lost through the high index bridges between the holes. This effect, by contrast, is very much stronger for high order modes than it is for lower order modes. As shown in Fig. 3.7, the microstructure can effectively act as a "sieve" confining modes which are spatially too large to pass through the bridges to a far greater extent than those with smaller lobes. The degree of confinement loss can vary over several orders of magnitude for different modes.

Confinement loss in a microstructured cladding is a property that cannot be properly understood using a naive "homogenised" description. In single-mode fibres and bandgap fibres, its impact is mostly felt in terms of defining the air-fraction or number of rings of holes that are required to reduce loss to the desired levels. In multimode fibres, it is particularly significant as the large differential mode attenuation effectively defines the modes that propagate down the fibre. It is unfortunate then that the calculation of confinement loss cannot be trivially incorporated into the standard software used for modelling optical fibres. For those who wish to work only with standard single-mode designs, the results are well established (Fig. 3.8) and illustrate the exponential decrease in confinement loss with an increase in the number of rings. For more unusual structures or multimode fibres, however, the modelling required for accurate loss determination can be computationally extremely demanding. Options are discussed in much greater detail in Chapter 4.

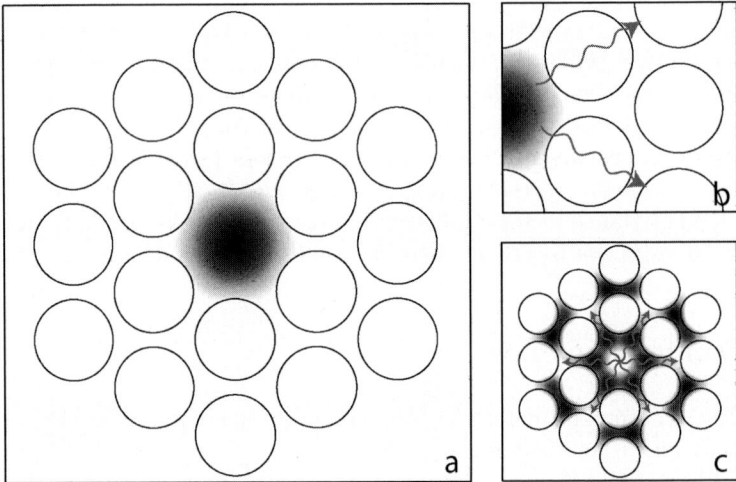

Fig. 3.7. The fundamental mode in a MOF is well confined by the hole structure (a), though some light tunnels through the air holes (b), and through the bridges between the holes (c). The latter process strongly affects the confinement of higher order modes whose smaller features are less confined by the hole structure.

3.2.1 Counting Modes In Microstructured Fibres

Since confinement loss determines which modes can propagate for any significant distance, let us now re-explore how one might count modes in microstructured fibres. The simple concept of cutoff fails for microstructured fibres: if there are no dopants, the index profile will have $n(r) \leq n_{cl}$ everywhere and thus using the definitions from Chapter 2, the fibre has no bound modes at all.

A pragmatic attempt [Argyros 2002] to count the number of modes that are "essentially" bound (i.e. have very low loss) defines the threshold for acceptable confinement loss directly in terms of the particular application for

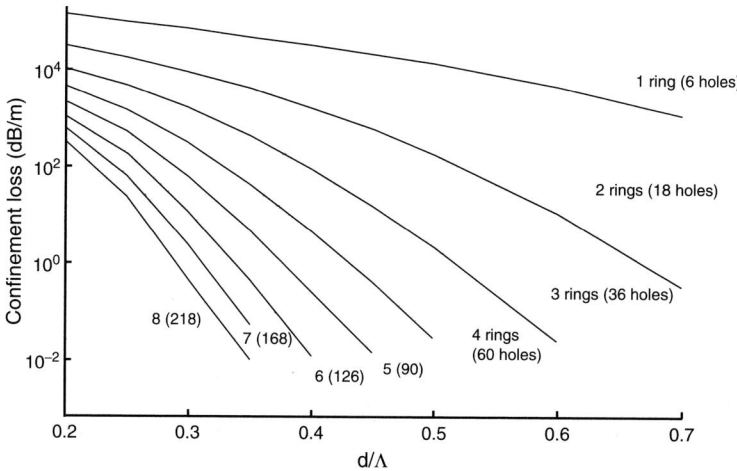

Fig. 3.8. The confinement loss of a single-mode microstructured fibre showing how the degree of leakage through the microstructure is dependent on the air-fraction of the cladding and the number of rings. The air-fraction is related to the ratio of hole diameter to hole spacing d/Λ. The confinement loss can be made arbitrarily small by appropriate choice of the structural parameters. After White et al. [2001]

the fibre. Most fibre applications will have a practical length L_{eff} associated with the fibre: if sufficient power persists in the mode over this length, then the mode is considered bound. Thus "pragmatically" single-mode fibres are those where the first mode propagates with almost no loss over the effective length, whilst the second mode has essentially disappeared by the end of the fibre.

A slightly more rigorous approach starts by attempting to define an *effective* cladding index for finite microstructured fibres: a concept closely related to numerical aperture. In a conventional fibre, the cutoff index is the same as the cladding index so we could rewrite the formula for numerical aperture as,

$$NA = \sqrt{n_{\text{co}}^2 - n_{\text{cutoff}}^2}, \tag{3.1}$$

thus, giving a direct relationship between cutoff and numerical aperture. Empirical evidence and recent analysis [Issa 2004] suggests that a better fit to the exit cone for certain multimode microstructured fibres is obtained by replacing the cladding index with an effective cladding or cutoff index. This idea of working with the index where the confinement loss increases beyond some acceptable threshold is very appropriate with graded-index microstructured fibre designs (see Chapter 9) were the transition from low to high loss is sharply defined.

Next, we look at a different example which highlights the importance of using an effective cladding index.

3.2.2 Endlessly Single-Mode Fibre

Consider a fibre where the numerical aperture increases with wavelength in such a way as to offset the wavelength dependence in the denominator of Eq. 2.3 for V. Such a fibre would have the same value of V at all wavelengths: in other words, it would have the same mode count at all wavelengths. Choosing this constant V small enough yields a fibre that is single-moded at all wavelengths.

An example of such a fibre [Knight et al. 1996] is shown in Fig. 1.4. The fibre has a well defined core with an index n_{co} that only varies weakly with wavelength (because of material dispersion). On the other hand, the cladding may be regarded as an infinite array of holes. The cladding index should be replaced by an effective or average $\bar{n}_{cl}$ that is determined by how much of the light penetrates into the holes and how much stays in the solid material. The penetration of the light into the holes is shown schematically in Fig. 3.9. At small wavelengths $\bar{n}_{cl}$ is close to n_{co} (average is close to the material index) while at longer wavelengths $\bar{n}_{cl}$ is lower and closer to n_{air} (average is close to index of the holes). Thus the index difference or numerical aperture *increases* with wavelength sufficiently rapidly to maintain a value of V consistent with single-mode guidance at all wavelengths.

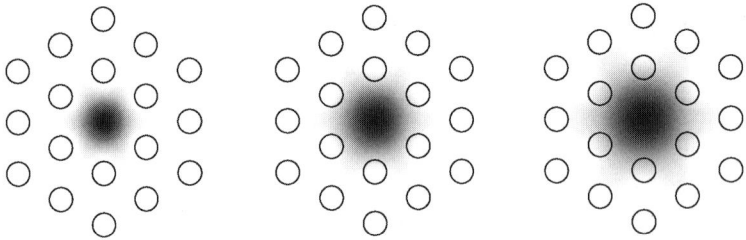

Fig. 3.9. The above schematics show how light penetrates into the cladding air holes for different wavelengths. The wavelength of light increases from left to right. The effective or average index of the cladding is reduced at longer wavelengths due to the field increasingly occupying the air regions.

Although this conceptual argument is adequate to qualitatively explain the phenomenon correctly, some assumptions may need further justification: such as replacing the cladding by an effective uniform medium when the wavelength is comparable to the scale of the microstructure. A more rigourous argument that gives better insight into the nature of the transition from modal confinement to nonconfinement was given by Kuhlmey et al. [2002] through two different mappings to effective conventional fibers that are valid asymptotically. Interestingly, the region between these asymptotic limits provides designs that have characteristics unique to microstructured fibres. This paper established the conditions necessary for the absence of a second mode cutoff, establishing that these fibres have at least one mode. A subsequent paper [Wilcox et al. 2005] established the absence of a fundamental mode cutoff; proving that these fibres do not have less than one mode. Thus, these two results in conjunction rigourously prove the single mode property. This allows the fibre to satisfy the single-mode condition for all wavelengths, and, in principle, for all core diameters.

Since the numerical aperture offsets the wavelength dependence in the expression for V this leaves the designer some latitude in the choice of core size. Single mode fibres with large core areas have been made, and this was initially highlighted as an important benefit of MOFs [Knight et al. 1998]. However, it has since become clear that large core single mode fibres are extremely sensitive to bend losses.

Endlessly single-mode fibres are merely one example of how MOFs exploit strong wavelength dependent or dispersive effects; many others are possible.

3.3 Manipulating Dispersion

Interest in the dispersion properties in MOFs arises from the ability to affect the waveguide component of the dispersion by changing the microstructure. Not only can this be manipulated in unusual ways, it also allows a measure of compensation for the material dispersion. This may be particularly significant in polymers where material dispersion is often higher than it is for silica [see Fig. 3.10(a)] particularly in the visible region where their transparency is highest. This has traditionally defined the limit of bandwidth performance of POF.

Figure 3.10(b) shows the dispersion curve (dispersion vs. wavelength) of a series of single-mode MOFs, with slightly different cladding structures, modelled using PMMA as the matrix material. The material dispersion of bulk PMMA is shown by the thickest curve. The dispersion curve of the MOF exhibits two wavelengths where the dispersion is zero, and a region of anomalous dispersion (where the short wavelengths travel faster than the long wavelengths). The zero dispersion points shift to shorter wavelengths as the size of the structure is reduced [Leon-Saval et al. 2005]. This means that it becomes straightforward to shift the zero group velocity dispersion point so that it coincides with a particular wavelength. Short pulses at this wavelength can then generate strong nonlinear effects because of the phase matching of different frequencies. One of the most important applications of MOFs to date has used exactly this effect to produce an extremely broad and bright white light source: a cascading series of nonlinear effects results in a dramatic spectral broadening known as *supercontinuum generation* [Ranka et al. 2000, Leon-Saval et al. 2004].

Very high intensity optical pulses are also used to produce other nonlinear effects in fibres. Physically these correspond to multiple photon interactions,

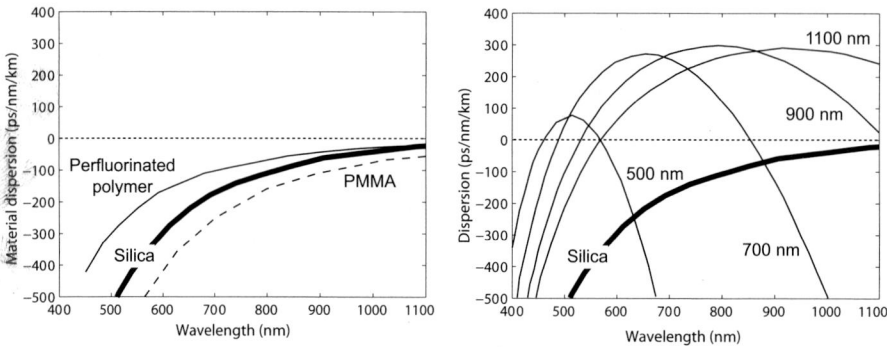

Fig. 3.10. The dispersion of a MOF can be widely varied by changing the microstructure. The material dispersion of perfluorinated polymer, silica and PMMA is shown in (a). The dispersion of silica MOFs of the same design but drawn to different diameters (as indicated) is shown in (b). For comparison, the material dispersion is also included. Data for (a) is reproduced from Koike and Ishigure [1999], while (b) is taken from Leon-Saval et al. [2004].

or the incident light interacting with an external electric field. These interactions transiently change the optical properties of the material, causing processes such as second harmonic generation, in which the energy of two identical photons is combined to form a photon with twice the energy, stimulated scattering processes such as Raman and Brillouin scattering, and pulse compression due to an intensity-dependent change in the refractive index. Applications which exploit these effects include the transmission of solitons, pulse shaping, wavelength conversion, optical data regeneration, optical demultiplexing, and Raman amplification.

In addition to these nonlinear optical applications, dispersion in MOFs has also been exploited in graded-index fibres for high-speed data transmission. In multimode fibres, there is an additional contribution to dispersion due to the different propagation constants of the modes. This is referred to as intermodal dispersion. The use of microstructures to control such intermodal dispersion is considered in Chapter 9.

References

Abeeluck, A K, Litchinitser, A N, Headley, C, and Eggleton, B (2002). Analysis of spectral characteristics of photonic bandgap waveguides. *Optics Express*, 10(23):1320–33.

Argyros, A (2002). Guided modes and loss in Bragg fibre. *Optics Express*, 10(24):1411–7.

Argyros, A, Bassett, I M, van Eijkelenborg, M A, and Large, M C J (2004). Analysis of ring-structured Bragg fibres for TE mode guidance. *Optics Express*, 12(12):2688–98.

Argyros, A, Birks, T A, Leon-Saval, S G, Cordeiro, C M B, Luan, F, and Russell, P St J (2005a). Photonic bandgap with an index step of one percent. *Optics Express*, 13(1):309–14.

Argyros, A, Birks, T A, Leon-Saval, S G, Cordeiro, C M B, and Russell, P St J (2005b). Guidance properties of low-contrast photonic bandgap fibres. *Optics Express*, 13(7):2503–11.

Argyros, A, van Eijkelenborg, M A, Large, M C J, and Bassett, I M (2006). Hollow-core microstructured polymer optical fibers. *Optics Letters*, 31(2):172–4.

Issa, N A (2004). High numerical aperture in multimode microstructured optical fibers. *Applied Optics*, 43(33):6191–7.

Knight, J C, Birks, T A, Cregan, R F, Russell, P S, and de Sandro, J P (1998). Large mode area photonic crystal fibre. *Electronics Letters*, 34(13):1347–8.

Knight, J C, Birks, T A, Russell, P St J, and Atkin, D M (1996). All-silica single mode optical fiber with photonic crystal cladding. *Optics Letters*, 21(19):1547–9.

Koike, Y and Ishigure, T (1999). Bandwidth and transmission distance achieved by POF. *IEICE Transactions on Communications*, E82-B:1287–1295.

Kuhlmey, B T, McPhedran, R C, and de Sterke, C M (2002). Modal cutoff in microstructured optical fibers. *Optics Letters*, 27(19):1684–6.

Leon-Saval, S G, Birks, T A, Joly, N Y, George, A K, Wadsworth, W J, Kakarantzas, G, and Russell, P S J (2005). Splice-free interfacing of photonic crystal fibers. *Optics Letters*, 30(13):1629–31.

Leon-Saval, S G, Birks, T A, Wadsworth, W J, and Russell, P St J (2004). Supercontinuum generation in submicron fibre waveguides. *Optics Express*, 12:2864–2869.

Litchinitser, N M, Abeeluck, A K, Headley, C, and Eggleton, B J (2002). Antiresonant reflecting photonic crystal optical waveguides. *Optics Letters*, 27(18):1592–4.

Parker, A R, McPhedran, R C, McKenzie, D R, Botten, L C, and Nicorovici, N P (2001). Photonic engineering: Aphrodite's iridescence. *Nature*, 409:36–7.

Ranka, J K, Windeler, R S, and Stentz, A J (2000). Visible continuum generation in air-silica microstructure optical fibers with anomalous dispersion at 800 nm. *Optics Letters*, 25(1):25–7.

Vienne, G, Xu, Y, Jakobsen, C, Deyerl, H-J, Jensen, J, Sørensen, T, Hansen, T, Huang, Y, Terrel, M, Lee, R, Mortensen, N, Broeng, J, Simonsen, H, Bjarklev, A, and Yariv, A (2004). Ultra-large bandwidth hollow-core guiding in all-silica Bragg fibers with nano-supports. *Optics Express*, 12(15):3500–8.

White, T P, McPhedran, R C, de Sterke, C M, Botten, L C, and Steel, M J (2001). Confinement losses in microstructured optical fibers. *Optics Letters*, 26(21):1660–2.

Wilcox, S, Botten, L, de Sterke, C M, Kuhlmey, B, McPhedran, R, Fussell, D, and Tomljenovic-Hanic, S (2005). Long wavelength behavior of the fundamental mode in microstructured optical fibers. *Optics Express*, 13:1978–84.

4

The Modelling and Design
of Microstructured Polymer
Optical Fibres

*The world can doubtless never be well known by theory: practice is
absolutely necessary; but surely it is of great use to a young man, before
he sets out for that country, full of mazes, windings, and turnings, to
have at least a general map of it, made by some experienced traveller.*

Lord Chesterfield, 1694-1773 British Statesman and Author

The first part of this chapter is about algorithms for modelling microstructured fibres. We also begin by briefly summarising the two major conventions for naming modes. The ideas behind the algorithms for calculating modes are then discussed but detailed results available in the literature are not reproduced here. It is impossible to be comprehensive or even perfectly balanced when covering such a wide field. Some references to both commercial and free software are given. We also sometimes give examples of how these algorithms have been used to analyse interesting features of microstructured fibres.

The existence of so many algorithms is further evidence that there cannot be a general, universal design process since there are many different types of fibres and the design of each will have its own issues: fibres for short distance purposes (e.g. end planes, endoscopes, automotive) may concentrate on spot shape and size; those for longer distances may focus on dispersion and bend loss.

The second part of this chapter surveys designs from the existing literature. These are fibre designs where a combination of expert intuition and some degree of analysis has been used to determine the best geometry to use; or to understand the origin of problematic features and alter the design to reduce such effects. Since many design issues are not specific to either silica or polymer, examples for both types of fibres are included. These designs are explored thematically according to the properties most relevant to the design looking at:

- Number and spacing of modes
- light acceptance and coupling

- Mode shape and evanescent fields
- Dispersion
- Polarisation

The references to the literature are not intended to be comprehensive, but to give interested readers a useful starting point and a flavour of the vast variety of effects that can be achieved with different designs.

The third part of this chapter is a discussion of automated optimisation and design in situations where expert intuition and optimising a small number of parameters are insufficient. These situations include attempts to simultaneously optimise multiple (possibly conflicting) criteria or to incorporate difficult manufacturing constraints into the design procedure. For such microstructured designs it is usually impossible to point to a particular hole or bridge (or other geometric feature) and claim that it is responsible for one effect or another.

4.1 Nomenclature

The naming of optical fibre modes is based entirely on the modes of circularly symmetric fibres and the labels reveal the polarisation and symmetry of the mode. The extent to which these names can be applied to modes of microstructured fibres depends entirely on how similar those modes are to those of circular fibres.

There are two commonly used nomenclature systems: the first system [Snitzer 1961] uses the labels TE, TM, HE and EH and is based on the rotational symmetry of the *longitudinal* field components of the modes; the second system [Gloge 1971] uses the label LP and is really only appropriate for weakly guiding fibres and is based on the rotational symmetry of the *transverse* field components.

Vector Nomenclature

Although the transverse components of the electric and magnetic fields are usually much larger than the longitudinal fields, this nomenclature system is based on the magnitude and symmetries of the *longitudinal* electric and magnetic fields. See Fig. 4.1 for the polarisation patterns of each type of mode.

The label TE stands for *transverse electric* and, as the name implies, the longitudinal component of the electric field is *zero*. The label TM stands for *transverse magnetic* and, as the name implies, the longitudinal component of the magnetic field is *zero*. The labels HE and EH indicate that the mode has both electric and magnetic components in the longitudinal direction. In addition, the E_z and H_z components are in quadrature (differ in phase by $\pm\pi/2$). The order of the letters in HE or EH indicates which field component leads or lags by $\pi/2$.

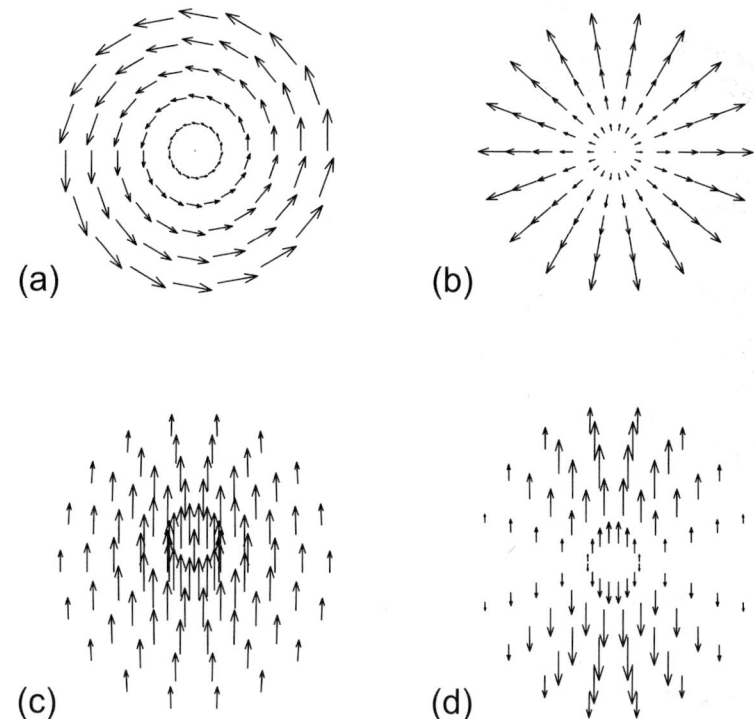

Fig. 4.1. Transverse polarisation for some common mode types. (a) The electric field for TE modes or the magnetic field for TM modes. (b) The electric field for TM modes or the magnetic field for TE modes. (c) The electric field for HE_{11} even or the magnetic field for HE_{11} odd (d) The electric field for EH_{11} even or the magnetic field for EH_{11} odd.

Each label is always followed by two subscripted integers, e.g. HE_{11} or TM_{02}. The choice of symbols for these subscripts in the standard reference by Snyder and Love [1983] is extremely unfortunate and thus will be avoided here. The first integer gives the azimuthal index which may be zero or positive and describes how the fields vary in the azimuthal direction (see Table 4.1). The second integer gives the radial index and starting from 1, basically counts the modes, with each higher-order mode having an additional node or oscillation in the transverse direction.

The azimuthal index (first subscript) for TE and TM modes is always zero, since the longitudinal components of these modes do not vary with angle. These modes are also always non-degenerate and described uniquely be their label. The HE and EH modes are always doubly degenerate, and can be decomposed into a pair of orthogonal polarisation states. The choice of states is not unique and the two most common alternatives are a pair of even and odd modes (with approximately horizontal and vertical polarisations) or a pair of left- and right-handed modes (with circular polarisation). Table 4.1 shows the angular dependence of the mode profiles for these degenerate modes.

Table 4.1. The differences between HE_{mn} and EH_{mn} modes as evidenced by the angular dependence of the longitudinal components. The integer m is always positive.

Parity	HE modes		EH modes	
	E_z	H_z	E_z	H_z
Even	$\cos(m\phi)$	$-\sin(m\phi)$	$\cos(m\phi)$	$\sin(m\phi)$
Odd	$\sin(m\phi)$	$\cos(m\phi)$	$\sin(m\phi)$	$-\cos(m\phi)$
Right Circular	$\exp(im\phi)$	$i\exp(im\phi)$	$\exp(im\phi)$	$-i\exp(im\phi)$
Left Circular	$\exp(-im\phi)$	$-i\exp(-im\phi)$	$\exp(-im\phi)$	$i\exp(-im\phi)$

Scalar Nomenclature

This nomenclature that uses the label LP is based on the behaviour of the *transverse* fields. One of the advantages of the weak-guidance approximation that has already been mentioned is that the *transverse* fields satisfy a scalar wave equation. Another is that the longitudinal fields have vanished in this approximation. The label LP reminds us that if we use cartesian components then in the weak guidance limit the transverse fields are all *linearly polarised*. This label also takes two integer subscripts which have the same meaning as in the vector notation, *except* based on the *transverse cartesian* field components.

For weakly guiding fibres, each scalar mode corresponds approximately to a group of *nearly degenerate* vector modes. Unfortunately, as Table 4.2 shows the two labelling systems do not match up in what one would have thought was the obvious way.

Table 4.2. Transverse polarisation for some common mode types in weakly guiding fibres. The integer subscripts are $n > 0$ and $\ell > 1$.

Scalar mode	Vector modes
LP_{0n}	HE_{1n}
LP_{1n}	$TE_{0n}, TM_{0n}, HE_{2n}$
$LP_{\ell,n}$	$HE_{\ell+1,n}, EH_{\ell-1,n}$

4.2 Algorithms

One would imagine after several decades that the numerical calculation of the modes of optical fibres would be routine, yet every year numerous new algorithms and numerical formulations appear in the literature. It has been a while since there was a systematic review of the field, and certainly the best entry points remain two reviews [Chiang 1994, Vassallo 1997]. A briefer review was given more recently [Scarmozzino et al. 2000]. The advent of microstructured fibres has only added to this collection of algorithms. Reviews specific to modelling microstructured fibres are also available [Bjarklev et al. 2003, Peyrilloux et al. 2002].

It is not too difficult to find the field profile and effective index of the fundamental mode of even a very complicated fibre. Most commercial software products will do a good job on these tasks, since almost any decent algorithmic approach can cope with fundamental modes. Many secondary quantities can then be calculated from the field profile: spot-size, nonlinear effective area, bend loss, group velocity. Accurate estimates of dispersion, especially when it is close to zero, require more precision in the effective index and this may need custom-built mode solvers. Commercial products are rarely designed with perturbations and imperfections in mind, and then one falls back on in-house programs or adapting existing algorithms. If you want to extract a trend or develop a useful semi-analytical approximation you are definitely left to your own devices.

As with other optical fibres, modelling MOFs involves solving Maxwell's equations for the appropriate fibre geometry, and with the correct boundary conditions. This process is carried out numerically, and yields the modes of the fibre, from which optical properties such as the birefringence or dispersion can be deduced. A wide variety of modelling techniques have been applied to MOFs, but each have advantages and disadvantages that make it hard to identify any one as a generically best technique. Depending on the application of interest, the reliability of the technique, its ability to calculate a particular property, the ease with which it can be automated and its robustness may motivate the choice of a particular technique. This chapter does not aim to do more than give an overview of some of the important issues and approaches to the problem.

4.2.1 Conventional Approaches

Finite difference and finite elements

Such methods both proceed by discretising the computational domain (a physical region which includes all of the microstructure) so that a matrix equation with a finite number of unknown quantities can be obtained. These methods are extremely powerful and allow the modal properties of arbitrary structures to be calculated, but are relatively computationally intensive. An issue that can arise in both approaches is the role of the symmetry of the mesh used to divide up the computational domain. Clearly, the results should be independent of the type of mesh, but in some cases the relationship of the mesh and fibre symmetry can cause anomalous results.

Finite difference methods replace all the differential operators with approximate discrete differences. Finite element methods divide the domain into sub-regions that are sufficiently small that the properties are uniform within the element. Maxwell's equations are then applied to each element, with continuity conditions applied at each boundary. Finite element methods are generally regarded as superior to finite difference (especially when derived rigorously from variational principles) but are more difficult to implement.

A full-vectorial finite-difference mode solver that uses standard meshing and an index averaging technique was successfully applied to microstructured fibres [Zhu and Brown 2002].

An implementation of the finite element approach is commercially available [www.comsol.com/products/electro] and widely used. Various applications of and modifications to finite element methods appear regularly in the literature. For example, a modification with curvilinear hybrid edge/nodal elements was used to explore both the single-mode nature of index-guiding MOFs as a function of wavelength and the effective index of the infinite photonic crystal cladding if the core is absent [Koshiba 2002]. More recently, a finite element method was also used to optimise the dispersion of microstructured fibre for use in an erbium-doped, hole-assisted optical fibre amplifier taking into account the dispersion of the germania/silica glass, the erbium emission and absorption cross-sections and the propagation loss [Prudenzano 2005].

Expansion methods

Another class of techniques expresses the solution to Maxwell's equation as a superposition of known functions. A particularly widely used approach here is the plane wave decomposition [Broeng et al. 1999]. Although this method can be used for arbitrary hole structures, it is most appropriate for truly periodic structures, where Bloch theory allows the solution of the wave equation to be expressed as a plane wave modulated by the periodicity of the crystal. Incorporating non-periodic features, which in the fibre case necessarily includes the core, requires a *supercell* to be defined, which includes the core, and is periodically repeated. A large number of terms are required to accurately represent these features, making the method computationally intensive. It has however, been widely used by the MOF community.

For example, recently the defect modes in two-dimensional photonic crystals with infinite claddings were analysed by representing the defect mode field as a superposition of solutions of quasi-periodic field problems [Wilcox et al. 2005]. The authors also simplified the two-dimensional superposition to a more efficient, one-dimensional average using Bloch mode methods.

Beam propagation method

This approach was not initially intended to find modes but rather to simulate propagation along non-uniform waveguides, tapers and couplers. The structure is divided into thin discrete layers transverse to the direction of propagation; then the effects of diffraction and refraction are alternately considered. Diffraction can be modelled either by finite differences or Fourier transform techniques leading to a diversity of beam propagation methods. This is a computationally intensive approach, which is probably more suited to non-uniform structures. A popular commercial version is available [www.rsoftdesign.com].

This method has been used to study the higher-order guided modes and so-called inner-cladding modes in a fibre with a ring of six large air-holes surrounding a doped core [Kerbage et al. 2000].

Finite difference time domain

Whereas all the methods discussed so far calculate the properties of the fibre at some fixed wavelength, the Finite Difference Time Domain (FDTD) method discretises the time dependent version of Maxwell's equations. Like the beam propagation methods, the equations are then solved in a leap-frog manner; alternating between the electric and magnetic fields at successive instants in time. The method is mostly used to simulate pulse propagation and problems involving reflection.

The band structure of a tapered transverse bandgap fibre has been studied using FDTD. Understanding how the bandgaps change with tapering may be used to monitor the draw process for bandgap fibres [Nguyen et al. 2004] or study how well fibre geometry is preserved during tapering [Domachuk et al. 2005].

4.2.2 Nonconventional Approaches

As discussed in both Chapters 2 and 3, one of the defining features of microstructured fibres is that their modes usually exhibit confinement loss. To a large extent, the diversity of methods available for modelling MOFs reflects the variety of ways in which confinement loss is modelled. Some of these approaches are summarised in Table 4.3.

Some of the methods in this table can be combined with the conventional algorithms already discussed: these are the methods that use an *artificial* layer beyond the hole structure to emulate the leaky nature of the microstructure. This layer can variously be absorbing, transparent or perfectly impedance matched. In the latter case, the artificial layer absorbs all the incident light without reflection. These artificial layer methods can work very well, but require skill in their application. The last two methods in the table are less conventional and do not rely on the introduction of artificial layers though both correctly model the confinement loss of the mode.

Table 4.3. Approaches used to address the leaky boundary conditions (BC) in MOFs.

Implementation	Advantages	Disadvantages
Absorbing layer	Automated mode search. Efficient for multimode fibres.	Spurious modes. Requires large computational domain to minimise back reflection.
Transparent BC	Highly automated mode search.	Requires iteration for each mode. Approximate.
Perfectly matched layer	Automated mode search. Efficient for multimode fibres.	Requires skillful parameter choice for each application. Spurious modes.
Multipole expansion	Most efficient for circular holes. No artificial boundary required.	Slow (sometimes manual) mode searches in the complex plane. Restricted hole shapes.
Adjustable BC	Highly automated mode search.	Requires iteration for each mode.

Multipole method

The multipole method [White et al. 2002] is based on an idea that dates back over a century to Lord Rayleigh [1892] where the field in the vicinity of a scattering object (in the case of MOFs, a hole) is separated into two parts: one part has the appropriate form for *outward* propagation, or behaviour at large distances; the other part is required to have the correct *inward* (or regular) behaviour at the origin of the scatterer. For example, the electric field around the i-th scatterer would be written

$$\mathbf{E}^i_{\text{tot}}(\mathbf{r}) = \mathbf{E}^i_{\text{in}}(\mathbf{r} - \mathbf{r}_i) + \mathbf{E}^i_{\text{out}}(\mathbf{r} - \mathbf{r}_i) \qquad (4.1)$$

where $\mathbf{E}^i_{\text{in}}(\mathbf{r} - \mathbf{r}_i)$ is required to be finite at $\mathbf{r} = \mathbf{r}_i$ and $\mathbf{E}^i_{\text{out}}(\mathbf{r} - \mathbf{r}_i)$ has the correct behaviour for large $\mathbf{r}$: evanescent for bound modes and outward radiating for leaky modes.

The geometric and material properties of the scatterer provide one link between these two parts via boundary conditions applied at the surface of the scatterer. For simple scatterers, such as circular holes, these boundary conditions lead to explicit expansions in terms of multipoles. In MOFs these multipoles are built from precisely the same types of Bessel functions (J_m, K_m and $H_m^{(1)}$) that are used to construct the modes of step-index circular fibres. For example, in cylindrical coordinates, the interior regular part of the longitudinal electric field looks like

$$E^i_{z,\text{in}}(r, \phi, z) = \sum_m a_m J_m(k_\perp^i r) e^{im\phi + ik_\parallel z} \qquad (4.2)$$

where $k_\perp$ and $k_\parallel$ are the transverse and longitudinal components of the wavevector, m is the order of the multipole, and a_m is an expansion coefficient. Similar expressions are obtained for the other field components. The outward part looks like

$$E^i_{z,\text{out}}(r, \phi) = \sum_m b_m K_m(\kappa_\perp r)e^{im\phi + ik_\parallel z} \tag{4.3}$$

for bound modes (where $\kappa_\perp$ is the transverse evanescent wavevector) and

$$E^i_{z,\text{out}}(r, \phi) = \sum_m b_m H^{(1)}_m(k_\perp r)e^{im\phi + ik_\parallel z} \tag{4.4}$$

for leaky modes. For circular scatterers, an explicit relationship between pairs of expansion coefficients a_m and b_m can be obtained in terms of the size and material properties of the scatterer. Note that the dependence on the azimuthal angle ϕ is in the form of a Fourier series. Exponential functions have been used here but one can also work with trigonometric Fourier series. However, much of the power and elegance of the multipole approach is lost when attempts are made to apply it to non-circular holes.

The second link between the two parts comes from the Rayleigh identity which states that the *inward* part of the field around any scatterer is equal to the superposition of the *outward* parts of the fields from all other scatterers:

$$\mathbf{E}^i_{\text{in}}(\mathbf{r} - \mathbf{r}_i) = \sum_{j \neq i} \mathbf{E}^j_{\text{out}}(\mathbf{r} - \mathbf{r}_j) \tag{4.5}$$

This identity can be exploited by taking the various local multipole expansions and rewriting them in a common coordinate system. Again, this step relies on the existence of explicit addition theorems [Abramowitz and Stegun 1964] for expansion functions.

These two links together lead to a matrix equation. The zeros of the determinant of the matrix yield the eigenvalues which are related to the effective indices of the modes and the corresponding eigenvectors yield the expansion coefficients a_m or b_m from which the field profile can be obtained.

Computationally this can be an extremely efficient method, and is analytically rigorous. An implementation of the method is available as freeware [www.physics.usyd.edu.au/cudos/mofsoftware]. Its disadvantages are that it is currently suitable only for circular holes, and finding the zeros of the determinant in the complex plane may be slow and require much human intervention.

Being a more analytical approach than some of the others, it can yield more profound physical insights. For example, the effect of a single scatterer or hole can be isolated from those of the environment [Kuhlmey 2005]. Various modifications of the basic approach have also been made. For example, the fibre structure can be decomposed into radial shells, allowing the confinement of light to be analysed layer by layer. This approach has suggested a number of new approaches to bandgap fibre design [Fini 2003].

Adjustable boundary condition method

This method [Poladian et al. 2002] is based on the idea of iterating to obtain ever more accurate estimates of the effective index. The method begins by choosing a radius R such that all the structure (holes, scatterers, dopants) are located inside that radius and there is only a uniform cladding outside. Thus, the method intrinsically cannot deal with structures that have an infinite number of holes. This enables the form of the field outside R to be written down analytically. Not surprisingly, the form of the field in the cladding is given by exactly the same expressions that are used in the outward parts of the fields in the multipole method: Eq. 4.3 for a bound mode and Eq. 4.4 for a leaky mode. The external expansion coefficients b_m are, of course, unknown and the various wavevectors ($k_\perp$, $\kappa_\perp$ and $k_\parallel$) depend on the effective index which is also unknown.

The method begins by making any reasonable initial guess of the effective index. In practice, the algorithm is extremely forgiving and no special knowledge or care is needed for this initial step. The effective index and the known form of the external fields are then used to determine what happens on the boundary $r = R$. The known external fields and their radial derivatives provide a set of boundary conditions that the unknown internal fields must satisfy. Because the boundary is chosen to to be a circle, the external expansion in multipoles is a Fourier series in the azimuthal angle ϕ. Further, the different terms in the external expansion are *not* coupled together at the boundary. This also means that the unknown external expansion coefficients b_m do *not* appear in the boundary conditions.

Like the multipole method, the elegance of the results so far arises from the choice of a circular boundary. However, unlike the multipole method the structure inside the circle of radius R can be completely arbitrary. Any computational method can be used that can solve Maxwell's equations inside a *finite* domain with prescribed conditions on an enclosing circular boundary. The initial choice was an expansion method using Fourier series in the azimuthal angle and Fourier-like series in the radial coordinate [Poladian et al. 2002, Issa and Poladian 2003]. Later, a hybrid method was chosen which retained the Fourier decomposition for the azimuthal variable but used a finite difference scheme in the radial direction [Issa 2005]. This resulted in considerable improvement in computational efficiency.

The effective indices obtained by solving the problem inside radius R can now be used to re-determine the external solutions and the entire calculation can be iterated. For most practical MOF designs the number of iterations required is about 3 to 5. The computational accuracy for each iteration is improved by choosing the radius R as small as possible, however the total number of iterations required is related to the fraction of the total power contained within the radius R so these two requirements need to be compromised. Again, for most practical MOF designs, the radius R is chosen to lie immediately beyond the outermost ring of holes.

Current limitations

A common problem with many algorithms is what can be referred to as mode-tracking error: when calculating the properties of a particular mode as a function of wavelength, the algorithm abruptly switches from one mode to a different one when the wavelength is slightly changed. Standard approaches to this problem are to change the wavelength very slowly and use algorithms where the mode closest (in effective index or profile shape) is determined. A more reliable way to guarantee that one has obtained a particular mode is to ensure it is orthogonal to all lower order modes. However, this is computationally expensive since for each mode all the modes of lower order need to be calculated or retained (though, in many common situations one wants the solutions for these lower order modes in any case). Human intervention (to check for duplicated or missing modes) is, of course, an alternative solution.

The same problem occurs when one wishes to guarantee that all modes within a certain range have been found and, if the program is running autonomously, the same mode is not calculated more than once. For example, the multipole method requires finding zeros in the complex plane: having found several zeros one is no wiser as to where the location of other zeros might be. One resorts to using heuristics or clever approximations [Kuhlmey 2005], or some sophisticated approach to counting zeros within contours in the complex plane.

On the other hand, the adjustable boundary condition method is searching for eigenvectors and one can insist that any new mode be orthogonal to all modes previously found. Unfortunately, although the orthogonality of different solutions could be used to control which mode is found, in current implementations, this process does not commute with the iteration required to adjust the boundary conditions [Poladian 2005] and thus a complete set of modes must be calculated at each iteration rather than just iterating the new mode in isolation.

4.2.3 Available Software

We conclude this section on algorithms with a short table of available software. These have been well developed for modelling conventional fibres and in some cases are available as commercial software (see Table 4.4).

4.3 Power Mixing Theory

The algorithms discussed so far deal with the properties of ideal fibres: those whose profiles do not change along the direction of propagation. Real fibres will also have imperfections such as impurities, rough surfaces or microscopic bends. These perturbations will cause light to couple between the different modes of the ideal fibre.

Table 4.4. Currently available software for modelling MOFs. The first three programs are commercial, while the second two are freeware.

Software	Confinement loss	Band diagrams	Beam propagation
RSOFT	Yes	Yes	Yes
www.rsoftdesign.com			
FEMLAB	Yes	No	Yes
www.comsol.com/products/electro			
Mad Max Optics	Yes	No	No
www.madmaxoptics.com			
CUDOS	Yes	No	No
www.physics.usyd.edu.au/			
cudos/mofsoftware			
MIT bands program	No	Yes	No
ab-initio.mit.edu/mbp			

In truly single-mode fibre, the end result of these perturbations is to couple light from the bound mode to radiation modes or to the backward propagating version of the same mode. Both of these effects will result in a simple attenuation of the mode. Thus no new interesting phenomena occur as a result of these perturbations.

Most single-mode fibres are actually doubly-degenerate and perturbations can cause coupling between the two different polarisation states. In conjunction with dispersion, this leads to a problem called polarisation mode dispersion (PMD) which is a problem in long distance communication.

The situation in multimode fibres is quite different. Here, perturbations lead to a redistribution of the power over all the modes of the fibre. Rather than being a problem, many multimode applications take advantage of this equilibration of the power distribution. To understand how this mode mixing occurs, we need to look at the differences between deterministic and stochastic perturbations.

4.3.1 Deterministic and Stochastic Coupling

Deterministic mode coupling is described by the following coupled mode equations

$$\frac{\partial a_\mu}{\partial z} = \sum_\nu K_{\mu,\nu}(z) a_\nu(z) e^{i(\beta_\mu - \beta_\nu)z} \qquad (4.6)$$

where μ and ν are mode indices ranging over all modes, $a_\mu(z)$ are the slowly varying complex mode amplitudes, and $K_{\mu,\nu}(z)$ is the coupling between mode μ and mode ν caused by some longitudinally varying perturbation. The exponential factor contains the difference $\Delta\beta = \beta_\mu - \beta_\nu$ between the propagation

constants of the two modes. The exponential is usually a more rapidly varying factor than the coefficients, and tends to suppress the amount of coupling
unless the coupling factor $K_{\mu,\nu}(z)$ varies on the same length scale. Thus deterministic coupling is often a resonance phenomenon: power flows back and
forth between the modes. The direction of power flow is determined by the
relative phase of the mode amplitudes.

There can also be a deterministic component to power attenuation. Material and confinement losses can be represented by the imaginary part of
the propagation constant; they are, thus, automatically included in the above
formalism. A power attenuation constant can be defined by $\alpha_\mu = 2\mathrm{Im}(\beta_\mu)$.

Stochastic mode coupling is described by a quite different set of equations
which we will refer to as 'mixing equations'

$$\frac{\partial P_\mu}{\partial z} = -\alpha_\mu P_\mu + \sum_\nu \kappa_{\mu,\nu}[P_\nu(z) - P_\mu(z)] \tag{4.7}$$

where $P_\mu(z)$ is the statistical ensemble average of the modal power and $\kappa_{\mu,\nu}$
is a mode mixing coefficient. The power attenuation coefficient α_μ contains
the deterministic losses mentioned above. In addition to coupling between
propagating modes, random perturbations can also cause power to couple
to radiation modes. Rather than include the radiation modes in the coupled
power equations, it is possible to treat them as an external *environment* and
then coupling to radiation gets modelled as a loss.

There are some major differences between deterministic and stochastic
mode coupling. There is no interference phenomenon in the mixing equations:
the complex amplitudes do not appear, only the powers. These equations do
not exhibit any resonance phenomena and thus the power flow is not back and
forth. In fact, the power flow moves irreversibly towards some equilibrium distribution which is jointly determined by the mode mixing and the differential
mode attenuation.

The connection between the deterministic $K_{\mu,\nu}(z)$ coupling coefficient and
the stochastic $\kappa_{\mu,\nu}$ is given by

$$\kappa_{\mu,\nu} = \int_{-\infty}^{\infty} \langle K_{\mu,\nu}(z')^* K_{\mu,\nu}(z'-z) \rangle e^{i(\beta_\mu - \beta_\nu)z} dz \tag{4.8}$$

where the angled brackets indicate an ensemble average. The quantity within
the ensemble brackets is the two-point correlation function. Thus, the mode
mixing is related to the power spectrum of the correlations in the perturbation.
Mode mixing is strong when the power spectrum contains variations at spatial
frequencies corresponding to differences in the propagation constants.

A simple model for random perturbations is to assume a Gaussian correlation

$$\langle K_{\mu,\nu}(z')^* K_{\mu,\nu}(z'-z) \rangle = h_{\mu,\nu}^2 \exp(-z^2/L_c^2) \tag{4.9}$$

where $h_{\mu,\nu}$ is the root-mean-squared amplitude of the coupling coefficients
and L_c is the correlation length. It is common to assume that the correlation

length is the same for all modes. For such a simple model, the power spectrum is obtained by a Fourier transform and yields

$$\kappa_{\mu,\nu} = h_{\mu,\nu}^2 \sqrt{\pi} L_c \exp(-\Delta\beta^2 L_c^2/4) \qquad (4.10)$$

The important observation here is that the exponential factor decays very strongly with the difference in propagation constants. Thus, if the correlation length is long, only very closely spaced modes are directly coupled by the perturbation. As the correlation length becomes shorter, modes further apart can be directly coupled. A common assumption in the early literature was to allow only *nearest-neighbour* coupling.

4.3.2 Comparing Two Common Perturbations

In conventional multimode fibre systems, the most important source of perturbation is considered to be microbending. A simple model [Marcuse 1974] for the mixing coefficients gives

$$\kappa_{\mu,\nu} = \{\frac{k_\mu^\perp k_\nu^\perp}{2kn_{cl}a\Delta\beta^2}\}^2 q_{rms}^2 \sqrt{\pi} L_c \exp(-\Delta\beta^2 L_c^2/4) \qquad (4.11)$$

where q_{rms} is the root-mean-square curvature, and a is the radius of the core. The numerator of the prefactor contains the transverse wavenumbers for each mode; as expected, modes closer to cutoff have larger transverse wavevectors and are more strongly affected by microbending. The denominator contains the difference in propagation constants; mixing due to microbending is certainly dominated by nearest-neighbour coupling. Although the expression above was derived for a conventional parabolic graded index fibre, it still provides clues to designing fibres with low microbend losses. There are some extra factors in the above formula related to selection rules, symmetry and degeneracy of modes that have been suppressed to make the essential physics clearer. The reader is referred to [Marcuse 1974] for all the details.

By way of contrast, an approximate model of mixing due to surface roughness [Poladian 2004] looks like

$$\kappa_{\mu,\nu} = \{\frac{k\Delta n^2}{4a}\}^2 h_{rms}^2 \sqrt{\pi} L_c \exp(-\Delta\beta^2 L_c^2/4) \sum_{holes} \langle(\mathbf{E}_\mu \cdot \mathbf{E}_\nu)^2\rangle \qquad (4.12)$$

where h_{rms} is the root-mean-square height of the roughness and Δn^2 is the difference in the square of the refractive indices at the rough interfaces. The last factor is a summation over all holes and contains the average taken around the perimeter of each hole of the relevant electric field intensities.

Both of the above types of perturbation also cause loss by coupling to radiation modes. The stochastic component of the loss α_μ will have expressions analogous to those above. An extremely good starting approximation is to take the above mode-to-mode mixing coefficients and to replace β_ν by the propagation constant of radiation in the cladding kn_{cl} and to replace the field profile by a uniform intensity.

4.3.3 Equilibrium Length And Power Distribution

The mode mixing equations (4.7) have several steady-state solutions. Each solution represents a power distribution and has an *effective* loss that applies to the entire distribution rather than to individual modes. One of these distributions is distinguished by having the smallest effective loss $\alpha_{\text{eff}}^{(0)}$: this is the *equilibrium* distribution. It is also the only solution where the power distribution is non-negative. Once equilibrium is established, the effective loss of this distribution will be the observed loss.

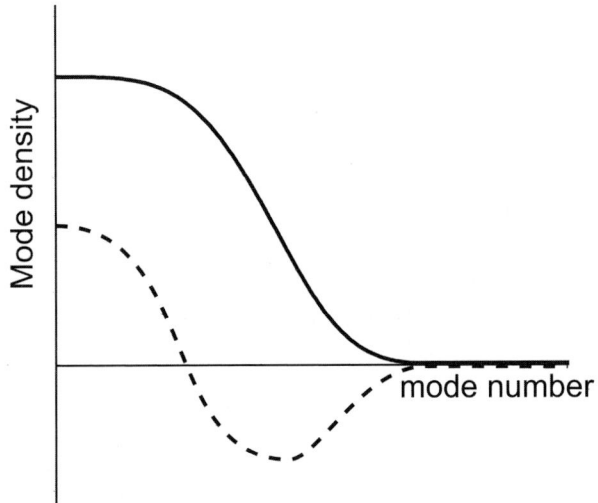

Fig. 4.2. A typical equilibrium power distribution is shown by the solid curve. The dashed curve shows another solution to the power mixing equations which represents the lowest order fluctuation from the equilibrium distribution. *Both* distributions are required to calculate equilibrium lengths and pulse broadening.

The other steady-state solutions have higher effective losses and represent power distributions that contain (non-physical) negative power. Thus, they cannot exist in isolation and represent *departures* from equilibrium. They are, however, extremely important as they describe the approach to equilibrium. The distribution with the *second* smallest effective loss $\alpha_{\text{eff}}^{(1)}$ determines the *equilibrium length*: it is the distance over which the least lossy distribution wins out over its closest competitor. The equilibrium length is

$$L_{\text{eq}} = \frac{1}{\alpha_{\text{eff}}^{(1)} - \alpha_{\text{eff}}^{(0)}} \tag{4.13}$$

This second distribution is also crucial in explaining pulse broadening. Once equilibrium is established the only source of pulse broadening will be fluctuations from equilibrium and, as discussed in Chapter 2, these are determined by a standard deviation leading to square-root length dependence. Because equilibrium pulse broadening is determined by fluctuations, its calculation requires determining the overlap of both the equilibrium and the *second* power distribution with the differential group delay of the fibre modes. This makes developing simple models of pulse broadening under equilibrium conditions quite challenging.

The application of mode mixing to graded index microstructured optical fibres (GImPOF) is presented and compared to experiment in Chapter 9.

4.4 Example Designs For Various Fibre Attributes

4.4.1 Mode Number And Spacing

Equivalent index designs

As discussed in Chapters 2 and 3, some useful insights can be obtained by simply modelling MOFs as a kind of traditional optical fibre, in which the presence of holes is used to modify the refractive index of the cladding. This equivalence can, in fact, be used to adjust the geometric parameters of the MOF to obtain desired optical properties, provided these do not rely on subtle interference or anti-resonance effects. This approach was successfully used to estimate the confinement loss of a MOF with a triangular air-hole lattice by replacing it with an equivalent doubly clad fibre with a depressed-index inner cladding [Koshiba and Saitoh 2005].

In a more sophisticated approach, a Bragg fibre is emulated by a MOF where the holes lie on circular rings. An average index model for these fibres assigns a lower index to each ring of holes. The size and spacing of the holes can then be chosen based on the equivalent Bragg fibre. Comparison to the calculated properties of the exact structure indicates that the ring structured MOFs behave approximately as the cylindrically layered fibres [Argyros et al. 2001].

Suppressing high-order modes

It is desirable to suppress high-order modes in many situations since such modes tend to have large bend losses or deleteriously affect the bandwidth of the fibre. One approach to suppressing these modes is to have them index-matched to various cladding modes [Fini 2005].

Mode spacing

The exact spacing of the effective indices is important in applications where gratings are written into the fibre. The grating period must match the mode spacing to provide resonant coupling. Experiments and simulations have shown that the mode spectrum (and thus the grating spectrum) is strongly affected by the design of the lattice of holes, but not so much by the cladding index [Eggleton et al. 1999, 2000]. Indeed, even the temperature dependence of the grating spectrum can be understood qualitatively and highly tunable long period gratings can be thus designed [Westbrook et al. 2000]. A tunable bandpass filter has also been designed using a variable thermal gradient to continuously control the transmission bands (i.e. wavelengths where mode coupling occurs) [Steinvurzel et al. 2005].

4.4.2 Capture And Coupling Efficiency

Large cores

Large cores are essential for ease of coupling (or for reducing nonlinear effects) and it is difficult to achieve large mode sizes using conventional or effective index designs. Most very large core designs rely on a bandgap guidance mechanism (as explained in Chapter 3). Usually the large core is required in addition to another fibre property. For example, a triangular lattice of high-index rods was used to achieve zero-dispersion in the near IR while maintaining an effective mode area of 17 μm^2 which is an order of magnitude larger than using conventional designs with similar dispersive properties [Riishede et al. 2004].

Large numerical aperture

Large *NA* designs are also essential for ease of coupling. Leaky modes have been used [Issa and Padden 2004] to determine the far-field angular intensity distributions which are subsequently used to calculate the capture efficiency and numerical aperture. Heuristic models were found to explain the dependence on length, wavelength, bridge thickness and number of layers of holes. Exceptionally high numerical apertures were found for structures where the bridge thicknesses was much smaller than the wavelength [Issa 2004].

Another application is in vertical cavity surface emitting lasers (VCSELs) achieving large-aperture single-mode emission. For example, hexagonal arrays of high-index cylinders in the upper mirror of a VCSEL [Liu et al. 2004].

4.4.3 Mode Profile

Designing fibres that have particular mode profiles is important for a variety of different purposes. Modes with a small spot-size or with an intensity profile which concentrates power near the core are useful for enhancing nonlinear effects. On the other hand, modes with long evanescent tails are useful for sensing. Fibres with low bend loss can be achieved by creating modes with fields that drop off rapidly in the transverse direction. Finally, non-circular modes are important for coupling between fibres and other devices such as sources and planar waveguides. Non-circular modes have also been shown to increase the pumping efficiency of fibre lasers.

Small spotsizes

Microstructured fibres with small-cores (relative to the wavelength) surrounded by a cladding with a large air fraction offer tight mode confinement and are therefore useful for high nonlinearity applications. For example, a recent optimised design in silica achieved effective nonlinearities of 52 $W^{-1}km^{-1}$ while still maintaining low confinement loss [Finazzi et al. 2003a,b].

Evanescent tails

The ability to design the geometry of microstructured fibres presents opportunities for a range of evanescent field devices for sensing applications. The existence of air holes can give access close to the fibre core and by introducing appropriate sensitive material into the air holes, a high interaction between light and the sensitive material can be obtained.

Design criteria for achieving significant overlap between the light guided in the fibre and the air holes have been developed [Monro et al. 2001]. The evanescent field in such structures can also be used for atom guiding [Noh and Jhe 2002]. Light modulation can also be achieved by incorporating tunable materials into the air holes. A variable attenuator (or loss filter) with a 30 dB dynamic range was formed using this method [Kerbage et al. 2001]. More recently, a liquid crystal filled photonic bandgap fibre has been used to create a thermo-optic fibre switch with an extinction ratio of 60 dB [Larsen et al. 2003].

Sensors that operate by allowing fluid entry into the microstructure holes can be designed to operate either by filling all the holes or just the central hole. Total internal reflection by the cladding microstructure may still provide robust confinement of light in a fluid-filled core if the average cladding index is sufficiently below the fluid index [Fini 2004].

Bending loss

In conventional fibres, the largest practical mode areas are limited by bending loss. The factors that influence bend loss in microstructured fibres may be quite different. The first detailed study of bend loss was able to correctly correlate the loss in large mode area fibres with experiment, and show that the hole configuration in the cladding has the strongest effect [Baggett et al. 2003].

Mode shape

In microstructured fibres, confinement is produced by the arrangement of the holes and thus it is relatively easy to obtain modes with unusual transverse shapes. One important application is to create a rectangular inner cladding to support the pump mode in fibre lasers. A Fabry-Perot ytterbium-doped laser was made using a double-clad MOF with a rectangular inner cladding [D'Orazio et al. 2005] and an mPOF of such geometry is presented in Chapter 7 Section 7.2.3. An alternative design for an erbium-doped fibre amplifier utilised an hexagonal shaped inner cladding has also been presented [Carlone et al. 2005].

4.4.4 Dispersion

Large index differences can be exploited to shift the zero dispersion wavelength to unusual parts of the spectrum, including into the very near IR and the visible. Tailoring the dispersion can allow many interesting nonlinear effects including the generation of a continuous broad spectrum extending from the violet to the infrared.

Zero dispersion wavelength

A microstructured fibre using band-gap guidance in a triangular lattice of high-index germanium-doped rods was used to move the zero-dispersion wavelength down to 730 nm, while maintaining a large effective mode area of 17 μm^2 [Riishede et al. 2004]. This area was an order of magnitude larger than what was previously achieved using index guidance for a similar zero-dispersion wavelength.

Moving the zero-dispersion wavelength also allows pulse compression and short pulse propagation to be explored at new wavelengths. For example, ten-fold compression of 150 fs pulses around 750 nm has been reported [Lako et al. 2003] using a microstructured optical fibre with zero-dispersion at 767 nm combined with a prism-pair/chirped-mirror compressor. Femtosecond soliton pulse propagation near the zero-dispersion point has also been investigated both experimentally and numerically [Fuerbach et al. 2005].

Fibres which have two separated zero-dispersion wavelengths have also been designed: for example, zero-dispersion points separated by more than 700 nm can help provide amplification at visible and infrared wavelengths [Genty et al. 2004].

Supercontinuum generation

Extending the design of the dispersion profile to obtain both zero dispersion and a very low dispersion slope creates fibres with broad regimes of very flat dispersion. This offers the possibility of creating ultrabroadband sources or supercontinuum generation. The spectral extent of the continuum is determined by the parametric gain curves for the modes involved in the nonlinear mixing about the zero-dispersion wavelengths of the relevant mode or modes [Dudley et al. 2002, Saitoh and Koshiba 2004]. Supercontinuum generation has been achieved with both LP_{01} and LP_{11} modes.

Dispersion compensation

Of course, the large index contrast in microstructured fibres can also be used to create extremely high dispersion which is useful for dispersion compensating fibres. For example, a fibre which has high dispersion but also matches the ratio of dispersion slope to dispersion has been designed [Zsigri et al. 2004] with dispersion exceeding 1350 psnm^{-1}km^{-1} at 1550 nm.

4.4.5 Polarisation

Any fibre design with a rotational symmetry of order three or higher cannot be birefringent. Modes which are predominantly linearly polarised always come in degenerate pairs. Modes which have no preferred axis of polarisation like TE and TM modes (and their approximate analogs in microstructured fibres) can be non-degenerate.

Usually in what are called 'single-mode' fibres (both conventional and microstructured), the relevant propagating mode is a doubly-degenerate pair; however, an air-core microstructured fibre design that supports a single-polarisation, circularly symmetric non-degenerate mode has been designed [Argyros et al. 2004]. Although this mode is the one with the smallest loss it is not the one with the largest propagation constant.

Many of the newer algorithms discussed earlier automatically incorporate group-theoretical results about mode degeneracy, and exploiting such symmetries leads to computational storage and time advantages. Surprisingly, inaccuracies in a few early numerical results led some to doubt these group-theoretical results. However, some early finite element methods were accurate enough [Koshiba and Saitoh 2001] to confirm that fibres with hexagonal symmetry are not birefringent and relevant pairs of modes are degenerate, in accordance with theory.

Microstructured fibres can be designed to manipulate polarisation by either locating the holes so as to break the symmetry for polarisation degeneracy, or by making individual holes that are non-circular, thus also breaking this symmetry (see Fig. 4.3).

For example, a fibre with an elliptical cross-section air holes was designed to be polarisation-preserving [Mogilevtsev et al. 2001]. The first fabrication of a microstructured optical fibre with uniformly oriented elliptical holes was achieved by exploiting hole deformation during the fibre draw [Issa et al. 2004]. Measurements of both form and stress-optic birefringence agreed with

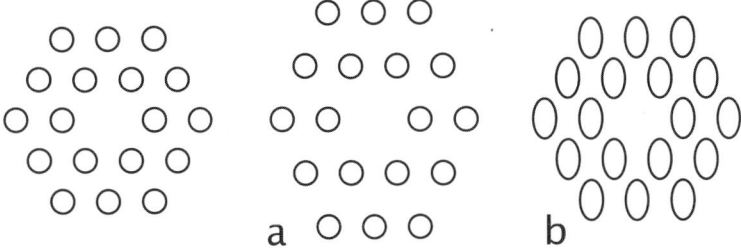

Fig. 4.3. Geometric birefringence can be introduced into a symmetric structure by (a) changing the locations of circular holes or (b) using elliptically shaped holes.

numerical modelling and demonstrated a birefringence as high as 1.0×10^{-4} at a wavelength of 850 nm (see Chapter 7 Sec. 7.2.7).

The degree of birefringence of a microstructured fibre can be tuned by filling the holes with an index tunable material, such as certain polymers [Kerbage and Eggleton 2002]. One can even start with a symmetric design and achieve asymmetry by selective filling of appropriately located holes [Kerbage et al. 2002].

4.5 Automated Design

Many authors use the terms 'design' and 'optimisation' interchangeably; in this section it is worth making a distinction. The word 'optimisation' should be interpreted as an activity where the parameters of a well established design or concept is altered to maximise (or minimise) various relevant attributes. Such a procedure can usually be automated because even though we do not know the exact answer in advance, we know most of its properties and can predict how our algorithms will behave. The various candidate design options will be qualitatively similar and we do not expect to learn anything new or have our intuition about the way the structure behaves invalidated.

On the other hand, 'design' in the greater sense takes place over a much broader range of alternatives. Although the problem is well defined and fixed, we do not have a pre-existing notion of the nature of the best option. Candidate solutions can be qualitatively different (in symmetry, topology, complexity etc.) This puts much more pressure on any numerical algorithms that we wish to incorporate into our search. Despite the sophistication of many available algorithms, the ability for these codes to consistently perform accurately over a vast array of designs without human intervention is still not adequate for many real-world problems (at this point in time). However, the frameworks within which future algorithms will be imbedded are within our grasp.

Another difference between optimisation and design is the way the space of possible alternatives is explored. Optimisation can be performed systematically by sequentially examining the total range of all parameters, or by using more sophisticated local search algorithms such as hill-climbing, conjugate gradient, or simulated annealing [Michalewicz and Fogel 2004]. Many design approaches, on the other hand, maintain a large population of candidate designs at all times and attempt to uncover global information about the space of all possible structures. This is useful if the fitness landscape (the way the desired properties vary over the space of designs) has a complicated topography with many local optima separated by deep barriers. Population-based searches can also exploit parallel computing.

Another, independent, reason for wanting to study a population of candidate designs is that real-world problems usually have many conflicting requirements. The relative importance of these objectives (or even which are easy or hard to achieve) may not be known in advance. This necessitates a multi-objective optimisation approach [Deb 2001, Coello and Lamont 2004]. In such an approach, the final solution is not a single design but a hyper-surface of designs in the space of all possible structures.

Evolutionary algorithms

Of the various approaches to using populations of candidate designs, the most common are those inspired by biological evolution. A population of candidate solutions (called individuals or phenotypes) are represented by a set of parameters (called the genotype). The genotype is either represented by a set of real variables or by a string of 0s and 1s, but other representations are also used. One starts with a population of randomly generated individuals. In each generation, the fitness of each individual in the population is determined. Here the fitness can be any suitable figure-of-merit related to the desired performance characteristics of the optical fibre. A subset of individuals are then selected (according to their fitness) from the current population. They undergo mutation and/or recombination to form the population for the next generation. Mutation consists of either flipping some bits in the binary string or adding perturbations to real-valued representations. Recombination takes two (or more) individuals called parents and creates one (or more) new individuals called children. The process can be as simple as inheriting some random subset of parameters from one parent and the remaining parameters from the other (this is common in the subset of algorithms called evolutionary strategies [Schwefel 1995]), or quite complicated cutting and splicing of binary strings (as is common in the subset of algorithms called genetic algorithms [Holland 1995]). Unsurprisingly, there are many variations to the type and order of selection, mutation, and recombination steps.

An early example [Krug et al. 2000] used only mutation and selection to develop non-zero dispersion-shifted fibres using designs that had various combinations of multiple cores and depressed claddings. The design space had from 6 to 10 degrees of freedom. The selection was based on the calculated dispersion at selected wavelengths and was partly automated and partly relied on human intervention to select those designs that would be fabricated and used to seed the starting population of the next iteration. Using this approach, several hundred thousand designs were explored numerically but less than a dozen were fabricated with each fabricated design (and subsequent population of evolved designs) being superior to those before.

The first published example of an evolutionary algorithm for *microstructured* fibres [van Eijkelenborg et al. 2001] used a relatively simple design space: the MOF consisted of rings of holes. The variables were the radius of the rings, the number of holes and the sizes of the holes. A number of characteristics such as splice loss, dispersion and non-linear effective area were aggregated into a single fitness parameter.

A comparison of an evolutionary strategy, simulated annealing, and the simplex algorithm [Manos et al. 2002] was applied to the optimisation of a hexagonal array MOF. The desired characteristics were minimal mode field diameter and maximum dispersion slope for supercontinuum generation. The design had fixed complexity (with 4 rings of holes) while the hole spacing and hole size formed the two parameters of the search space.

In contrast, designs with minimal dispersion have also been explored. A genetic algorithm was used [Kerrinckx et al. 2004] with a fixed number of rings and a low dimensional search space. The same problem was subsequently addressed [Poletti et al. 2005] using a much larger search space of 6 parameters by allowing the holes in each ring to have different sizes.

All the above examples share a common feature: the complexity of the design is essentially fixed. Indeed, in most cases, the qualitative aspects of the design are also imposed in advance and the evolutionary algorithms are merely optimising a set of parameters. However, the vast space of possible designs accessible by microstructured fibres has stimulated the development of a genetic algorithm with a more complex representation so that even the qualitative nature of the design need not be specified in advance.

The need for good recombination operators goes hand-in-hand with more complex representations. According to the building block hypothesis, recombination is the most important operation because it allows partial solutions that exist in the population to be brought together and thus accelerate the search for optimal solutions [Goldberg 1989].

Manufacturing constraints

One of the difficulties of allowing genetic algorithms to work with designs of arbitrary complexity is to simultaneously explore a large number of alternatives and ensure that all these designs are feasible (i.e. can be manufactured) or at least to minimise the amount of effort spent exploring the infeasible regions of the design space. The different approaches to handling infeasible designs include: eliminating them from the population, retaining the designs but assigning them a low fitness, or repairing the design by changing parameters to force it to be feasible. The first two approaches do not address the issue of wasting time generating and exploring infeasible solutions and unless the repair mechanism is sophisticated it tends to produce lots of designs clustered around the extreme limits of what can be fabricated (even if these are not genuinely active constraints). A more powerful approach (in situations where it is possible) is to use a representation and corresponding recombination operators that faithfully create and reproduce only feasible designs.

Again, the inspiration comes from biology in the developmental processes that transform a single cell into an embryo and finally an adult organism. An artificial embryogeny was developed for MPOFs [Manos et al. 2005] that incorporated the current fabrication constraints: holes are only available in a discrete set of sizes (corresponding to available drill bits) and the separation of neighbouring holes must be larger than a specified tolerance (to avoid hole wall breakage). The genotype encodes the location of each hole and its *potential* size. The rotational symmetry of the fibre is also a parameter encoded in the genotype. The embryonic fibre design begins with all holes having zero radius. Each hole increases through the available sizes until it either reaches its full potential size or it prematurely stops growing because it would otherwise be too close to another hole. Note that it is possible that some holes may not ever grow. Thus, the final design always satisfies the fabrication constraints. The recombination process produces children that inherit their hole positions and potential sizes from either parent. The embryogeny can then grow children that look quite different from either parents. Examples of mPOF designs that satisfy relevant manufacturing constraints can be seen in Chapter 9.

Multiobjective optimisation

The final issue related to automated design concerns the desired characteristics of optical fibres. It is extremely rare to only want to specify or optimise a single characteristic. Often many quite different attributes such as connectivity, loss and dispersion need to be considered simultaneously to produce a good fibre design.

The simplest approach to deal with multiple objectives is to aggregate them into a single figure of merit. This approach presupposes that the relative importance of the criteria is specified in advance. It is not unusual to then find, that having gone through a design process and seen some results and what range of designs are possible, that the end-user may change their mind about the relative weighting of objectives.

A different approach uses Pareto optimisation [Deb 2001]. A design is said to be dominated by another design if it is inferior with respect to *all* criteria. Starting with a very large population of designs, all dominated designs are deleted. The remainder is called the non-dominated set. If the original population consisted of *all* feasible designs, then its non-dominated set is the Pareto set. Evolutionary algorithms can be constructed [Deb 2001] such that with each passing generation, the non-dominated set yields successively better approximations to the Pareto set.

The Pareto set can be regarded as the set of all possible compromises between the various criteria of interest. Each design in the Pareto set chooses a different balance of objectives. The structure (i.e. dimensionality and topology) of the Pareto set can provide insight into how the different objectives depend on the various design parameters and allows for a more informed choice of the final choice of design compromise or aggregated figure of merit. Examples of mPOF designs that achieve multiple objectives are discussed in Chapter 9.

References

Abramowitz, Milton and Stegun, Irene A. (1964). *Handbook of Mathematical Functions with Formulas, Graphs, and Mathematical Tables.* Dover, New York, 9th Dover printing, 10th GPO printing edition.

Argyros, A, Bassett, I M, van Eijkelenborg, M A, Large, M C J, Zagari, J, Nicorovici, N A P, McPhedran, R C, and de Sterke, C M (2001). Ring structures in microstructured polymer optical fibres. *Optics Express*, 9(13):813–20.

Argyros, A, Issa, N A, Bassett, I M, and van Eijkelenborg, M A (2004). Microstructured optical fibres for single-polarisation air-guidance. *Optics Letters*, 29(1):20–3.

Baggett, J C, Monro, T M, Furusawa, K, Finazzi, V, and Richardson, D J (2003). Understanding bending losses in holey optical fibers. *Optics Communications*, 227(4-6):317–335.

Bjarklev, A, Broeng, J, and Bjarklev, A S (2003). *Photonic crystal fibres.* Kluwer academic publishers, Boston, USA.

Broeng, J, Mogilevstev, D, Barkou, S E, and Bjarklev, A (1999). Photonic crystal fibers: A new class of optical waveguides. *Optical Fiber Technology*, 5(3):305–330.

Carlone, G, amd M De Sario, A D'Orazio, amd V Petruzzelli, L Mescia, and Prudenzano, F (2005). Design of double-clad erbium-doped holey fiber amplifier. *Journal of Non-crystalline Solids*, 351(21-23):1840–1845.

Chiang, K S (1994). Review of numerical and approximate methods for the modal-analysis of general optical dielectric wave-guides. *Optical And Quantum Electronics*, 26(3):S113–S134.

Coello, C A Coello and Lamont, G B (2004). *Applications of Multi-Objective Evolutionary Algorithms.* World Scientific, Singapore.

Deb, Kalyanmoy (2001). *Multi-Objective Optimization using Evolutionary Algorithms.* John Wiley and Sons, Chichester.

Domachuk, P, Chapman, A, Magi, E, Steel, M J, and Nguyen, H C (2005). Transverse characterization of high air-fill fraction tapered photonic crystal fiber. *Applied Optics*, 44(19):3885–3892.

D'Orazio, A, de Sario, M, Mescia, L, Petruzzelli, V, and Prudenzano, F (2005). Design of double-clad ytterbium-doped microstructured fibre laser. *Applied Surface Science*, 248(1-4):499–502.

Dudley, J M, Provino, L, Grossard, N, Maillotte, H, Windeler, R S, Eggleton, B J, and Coen, S (2002). Supercontinuum generation in air-silica microstructured fibers with nanosecond and femtosecond pulse pumping. *Journal of the Optical Society of America B - Optical Physics*, 19(4):765–771.

Eggleton, B J, Westbrook, P S, White, C A, Kerbage, C, Windeler, R S, and Burdge, G L (2000). Cladding-mode-resonances in air-silica microstructure optical fibers. *Journal of Lightwave Technology*, 18(8):1084–1100.

Eggleton, B J, Westbrook, P S, Windeler, R S, and Spalter, S (1999). Grating resonances in air-silica microstructured optical fibers. *Optics Letters*, 24(21):1460–1462.

Finazzi, V, Monro, T M, and Richardson, D J (2003a). The role of confinement loss in highly nonlinear silica holey fibers. *IEEE Photonics Technology Letters*, 15(9):1246–1248.

Finazzi, V, Monro, T M, and Richardson, D J (2003b). Small-core silica holey fibers: nonlinearity and confinement loss trade-offs. *Journal of the Optical Society of America*, 20(7):1427–1436.

Fini, J M (2003). Analysis of microstructure optical fibers by radial scattering decomposition. *Optics Letters*, 28(12):992–994.

Fini, J M (2004). Microstructured fibres for optical sensing in gases and liquids. *Measurement Science and Technology*, 15(6):1120–8.

Fini, J M (2005). Design of solid and microstructure fibers for suppression of higher-order modes. *Optics Express*, 13(9):3477–3490.

Fuerbach, A, Steinvurzel, P, Bolger, J A, and Eggleton, B J (2005). Nonlinear pulse propagation at zero dispersion wavelength in anti-resonant photonic crystal fibers. *Optics Express*, 13(8):2977–2987.

Genty, G, Lehtonen, M, Ludvigsen, H, and Kaivola, M (2004). Enhanced bandwidth of supercontinuum generated in microstructured fibers. *Optics Express*, 12(15):3471–3480.

Gloge, D (1971). Weakly guiding fibers. *Applied Optics*, 10(10):2252.

Goldberg, D E (1989). *Genetic Algorithms in Search, Optimization, and Machine Learning*. Addison-Wesley, Reading, Masschusetts.

Holland, John H (1995). *Hidden Order, How Adaptation Builds Complexity*. Helix Books, Cambridge, Massachusetts.

Issa, N A (2004). High numerical aperture in multimode microstructured optical fibers. *Applied Optics*, 43(33):6191–6197.

Issa, N A (2005). *Modes and propagation in microstructured optical fibres*. PhD dissertation, The University of Sydney, Sydney, Australia.

Issa, N A and Padden, W E (2004). Light acceptance properties of multimode microstructured optical fibers: Impact of multiple layers. *Optics Express*, 12(14):3224–3235.

Issa, N A and Poladian, L (2003). Vector wave expansion method for leaky modes of microstructured optical fibers. *Journal of Lightwave Technology*, 21(4):1005–1012.

Issa, N A, van Eijkelenborg, M A, Fellew, M, Cox, F, Henry, G, and Large, M C J (2004). Fabrication and study of microstructured optical fibers with elliptical holes. *Optics Letters*, 29(12):1336–1338.

Kerbage, C and Eggleton, B J (2002). Numerical analysis and experimental design of tunable birefringence in microstructured optical fiber. *Optics Express*, 10(5):246–255.

Kerbage, C, Hale, A, Yablon, A, Windeler, R S, and Eggleton, B J (2001). Integrated all-fiber variable attenuator based on hybrid microstructure fiber. *Applied Physics Letters*, 79(19):3191–3193.

Kerbage, C, Steinvurzel, P, Reyes, P, Westbrook, P S, Windeler, R S, Hale, A, and Eggleton, B J (2002). Highly tunable birefringent microstructured optical fiber. *Optics Letters*, 27(10):842–844.

Kerbage, C E, Eggleton, B J, Westbrook, P S, and Windeler, R S (2000). Experimental and scalar beam propagation analysis of an air-silica microstructure fiber. *Optics Express*, 7(3):113–122.

Kerrinckx, Emmanuel, Bigot, Laurent, Douay, Marc, and Quiquempois, Yves (2004). Photonic crystal fiber design by means of a genetic algorithm. *Optics Express*, 12(9):1990–1995.

Koshiba, M (2002). Full-vector analysis of photonic crystal fibers using the finite element method. *IEICE Transactions on Electronics*, E85C(4):881–888.

Koshiba, M and Saitoh, K (2001). Numerical verification of degeneracy in hexagonal photonic crystal fibers. *IEEE Photonics Technology Letters*, 13(12):1313–1315.

Koshiba, M and Saitoh, K (2005). Simple evaluation of confinement losses in holey fibers. *Optics Communications*, 253(1-3):95–98.

Krug, P A, Poladian, L, and Large, M I (2000). Advanced fibre design by evolutionary computation. In *Proceedings of the 25th Australian Conference on Optical Fibre Technology*, pages 74–76.

Kuhlmey, B T (2005). Modelling microstructured optical fibres with the multipole method. In *14th International Workshop on Optical Waveguide Theory and Numerical Modelling*, page 18.

Lako, S, Seres, J, Apai, P, Balazs, J, Windeler, R S, and Szipocs, R (2003). Pulse compression of nanojoule pulses in the visible using microstructure optical fiber and dispersion compensation. *Applied Physics B - Lasers and Optics*, 76(3):267–275.

Larsen, T T, Bjarklev, A, Hermann, D S, and Broeng, J (2003). Optical devices based on liquid crystal photonic bandgap fibres. *Optics Express*, 11(20):2589–2596.

Liu, H R, Yan, M, Shum, P, Ghafouri-Shiraz, H, and Liu, D M (2004). Design and analysis of anti-resonant reflecting photonic crystal VCSEL lasers. *Optics Express*, 12(18):4269–4274.

Manos, S, Poladian, L, Bentley, P, and Large, M (2005). Photonic device design using multiobjective evolutionary algorithms. In *Lecture Notes In Computer Science : International Conference on Evolutionary Multi-criterion Optimization*, volume 3410, pages 636–650.

Manos, Steven, Mitchell, Arnan, Lech, Margaret, and Poladian, Leon (2002). Automated synthesis of microstructured holey optical fibres using numerical optimisation. In *Proceedings of the 27th Australian Conference on Optical Fibre Technology, 8th-11th July, Darling Harbour, Sydney, Australia*, pages 47–49.

Marcuse, D (1974). *Theory of dielectric optical waveguides*. Academic Press, New York.

Michalewicz, Zbigniew and Fogel, David B (2004). *How to Solve It: Modern Heuristics*. Springer.

Mogilevtsev, D, Broeng, J, Barkou, S E, and Bjarklev, A (2001). Design of polarization-preserving photonic crystal fibres with elliptical pores. *Journal of Optics A - Pure and Applied Optics*, 3(6):S141–S143.

Monro, T M, Belardi, W, Furusawa, K, Baggett, J C, Broderick, N G R, and Richardson, D J (2001). Sensing with microstructured optical fibres. *Measurement Science and Technology*, 12(7):854–858.

Nguyen, H C, Domachuk, P, Steel, M J, and Eggleton, B J (2004). Experimental and finite difference time domain technique characterization of transverse in-line photonic crystal fiber. *IEEE Photonics Technology Letters*, 16(8):1852–1854.

Noh, H R and Jhe, W (2002). Atom optics with hollow optical systems. *Physics Reports-Review Section of Physics Letters*, 372(3):269–317.

Peyrilloux, A, Fevrier, S, Marcou, J, Berthelot, L, Pagnoux, D, and Sansonetti, P (2002). Comparison between the finite element method, the localized function method and a novel equivalent averaged index method for modelling photonic crystal fibres. *Journal Of Optics A*, 4(3):257–262.

Poladian, L (2004). Modelling surface imperfection induced mode mixing in multimode microstructured fibres. In *Proceedings of the Conference on Lasers and Electro Optics*, San Francisco, USA.

Poladian, L (2005). Beyond computing leaky modes. In *14th International Workshop on Optical Waveguide Theory and Numerical Modelling*, page 18.

Poladian, L, Issa, N, and Monro, T (2002). Fourier decomposition algorithm for leaky modes of fibres with arbitrary genometry. *Optics Express*, 10(10):449–454.

Poletti, F, Finazzi, V, Monro, T M, Broderick, Tse, V, and Richardson, D J (2005). Inverse design and fabrication tolerances of ultra-flattened dispersion holey fibers. *Optics Ex[ress*, 13(10):3728–3736.

Prudenzano, F (2005). Erbium-doped hole-assisted optical fiber amplifier: Design and optimization. *Journal of Lightwave Technology*, 23(1):330–340.

Rayleigh, J W S (1892). On the influence of obstacles arranged in rectangular order upon the properties of a medium. *Philosophical Magazine*, 34:481–502.

Riishede, J, Laegsgaard, J, Broeng, J, and Bjarklev, A (2004). All-silica photonic bandgap fibre with zero dispersion and a large mode area at 730 nm. *Journal of Optics A - Pure and Applied Optics*, 6(7):667–670.

Saitoh, K and Koshiba, M (2004). Highly nonlinear dispersion-flattened photonic crystal fibers for supercontinuum generation in a telecommunication window. *Optics Express*, 12(10):2027–2032.

Scarmozzino, R, Gopinath, A, Pregla, R, and Helfert, S (2000). Numerical techniques for modeling guided-wave photonic devices. *IEEE Journal Of Selected Topics In Quantum Electronics*, 6(1):150–162.

Schwefel, H-P (1995). *Evolution and Optimum Seeking*. Wiley, New York.

Snitzer, E (1961). Cylindrical dielectric waveguide modes. *Journal Of The Optical Society Of America*, 51(5):491.

Snyder, A W and Love, J D (1983). *Optical waveguide theory*. Chapman and Hall, New York.

Steinvurzel, P, Eggleton, B J, de Sterke, C M, and Steel, M J (2005). Continuously tunable bandpass filtering using high-index inclusion microstructured optical fibre. *Electronics Letters*, 41(8):463–464.

van Eijkelenborg, M A, Zagari, J, and Poladian, L (2001). Optimising holey fibre designs. In *Proceedings of the OECC/IOCC Conference Incorporating ACOFT, 2-5th July*, pages 526–527.

Vassallo, C (1997). 1993-1995 Optical mode solvers. *Optical And Quantum Electronics*, 29(2):95–114.

Westbrook, P S, Eggleton, B J, Windeler, R S, Hale, A, Strasser, T A, and Burdge, G L (2000). Cladding-mode resonances in hybrid polymer-silica microstructured optical fiber gratings. *IEEE Photonics Technology Letters*, 12(5):495–497.

White, T P, Kuhlmey, B T, McPhedran, R C, Maystre, D, Renversez, G, Martijn de Sterke, C, and Botton, L C (2002). Multipole method for microstructured optical fibers i: Formulation. *Journal of the Optical Society of America B*, 19:2322–30.

Wilcox, S, Botten, L C, McPhedran, R C, Poulton, C G, and M de Sterke, C (2005). Modeling of defect modes in photonic crystals using the fictitious source superposition method. *Physical Review E*, 71(5).

Zhu, Z M and Brown, T G (2002). Full-vectorial finite-difference analysis of microstructured optical fibers. *Optics Express*, 10(17):853–864.

Zsigri, B, Laegsgaard, J, and Bjarklev, A (2004). A novel photonic crystal fibre design for dispersion compensation. *Journal of Optics A - Pure and Applied Optics*, 6(7):717–720.

5

Fabrication of Microstructured Polymer Optical Fibres

"Don't worry - we'll soon design the simplicity out of it."

A Royal Air Force Commander to Sir Frank Whittle

Conceptually, the fabrication of microstructured polymer optical fibre is a straightforward process involving just two steps - the creation of a preform containing a large-scale version of the structure desired in the fibre, followed by heating and drawing of the preform to produce the final fibre. Three issues soon emerge however when trying to turn this simple two-step process into a reality.

The first is how best to produce the desired microstructure within the preform. Most mPOFs to date have been made from polymethylmethacrylate (PMMA) which would generally be regarded as a well understood polymer. However even in this case a number of fabrication options exist. Thus far structured preforms for research purposes have been created by capillary stacking, by drilling into a monolithic rod, by extrusion, or by casting into a mould. Each of these fabrication techniques will be considered in this chapter, along with their suitability for commercial mPOF production.

The second is how best to heat and draw the structured preform to fibre in a controlled and reproducible way. The central problem is to provide sufficient heat to adequately soften a part of the preform. Two possible options are to use primarily convective or radiative heating. In this chapter, three draw towers will be considered, two using convective heating of the preform, while the other uses radiative heating. The importance of employing a short hot-zone where the preform necks down to fibre is discussed.

The final issue is concerned with how the overall process (i.e. both preform fabrication and drawing) impacts on the quality of the final fibre. Thus this chapter concludes with experimental data relating to fibre quality.

5.1 Preform Production

5.1.1 Bragg Preforms

Two methodologies have been used for fabrication of multilayered all-polymer Bragg fibre preforms [Gao et al. 2006]. One approach uses consecutive deposition of layers of two different polymers by solvent evaporation on the inside of a rotating polymer tube. Orthogonal solvents are required, and solvent-evaporation processes have been developed for two system pairs: Polymethyl methacrylate (PMMA) / Polystyrene (PS) and Polyvinylidene fluoride (PVDF) / Polycarbonate (PC). The other approach uses the 'swiss-roll' approach, rolling together two different polymer films. Examples of solid and hollow core Bragg preforms and fibres are shown in Fig. 5.1

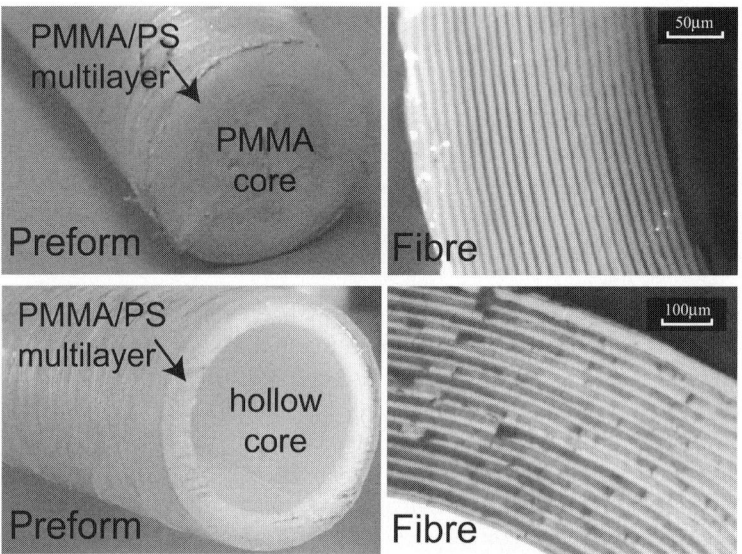

Fig. 5.1. Examples of all-polymer Bragg preforms and fibres. *Images courtesy of Maksim Skorobogatiy, École Polytechnique de Montréal.*

5.1.2 Capillary Stacking

The first examples of microstructured optical fibres were made from silica [Knight et al. 1996, Birks et al. 1997] with the preforms being made by hand using capillary stacking. A conventional draw tower was used to make capillaries with tightly controlled diameters which were then cut to length and stacked in the desired pattern [Knight et al. 2001].

This method has become the dominant method of fabricating silica MOF preforms. The capillary stack may be inserted into a tube to create a preform with the required structure, however the integrity of the stack requires the packing to be very tight to avoid excessive deformation during subsequent drawing to fibre. Figure 5.2 shows silica capillary stacked preforms with the central capillary replaced by a rod as a means of producing a solid core.

Fig. 5.2. Solid core silica preforms produced by capillary stacking both with and without an outer sleeve. *Image courtesy of Crystal Fibre A/S, Blokken 84, 3460 Birkerod, Denmark http://www.crystal-fibre.com*

There are many possible variations on the capillary stacking theme, and a wide range of preforms (and fibres) have been produced by using combinations of rods and capillaries of different diameters and/or different materials. For example, by stacking and drawing a capillary-based preform down to an intermediate "cane", it is possible to create a scaffold structure into which other fibres can be placed. This technique has been used [Argyros et al. 2005a,b] to produce the solid photonic bandgap fibre as shown in Fig. 5.3. A single-mode fibre was placed at the centre of the scaffold structure, to become the core of the bandgap fibre, while germanium-doped multimode fibres were placed around it, to become high index inclusions in the cladding.

Several groups have used capillary stacking to produce structured polymer preforms. In fact, owing to the large transverse dimensions, polymeric terahertz radiation waveguides have been fabricated directly (without drawing) by stacking either high density polyethylene [Han et al. 2002] or Teflon rods and tubes [Goto et al. 2004]. The first fibres made from capillary stacked PMMA preforms were reported several years ago [Park et al. 2002]. However residual stress in the (previously drawn) capillaries caused contraction of the

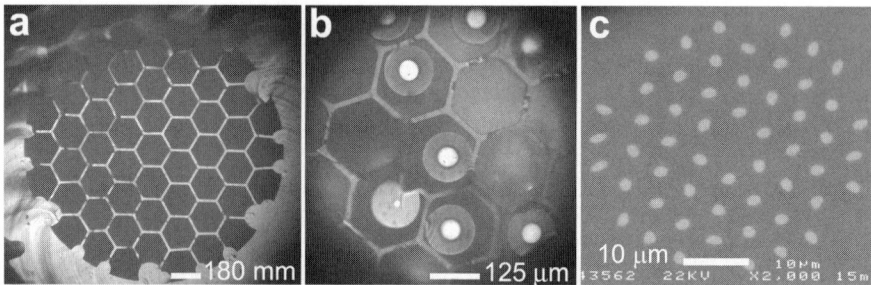

Fig. 5.3. (a) Stacked silica capillaries are drawn down to an intermediate cane stage to form a scaffold structure. (b) Scaffold is filled with conventional fibres. (c) Filled scaffold structure is drawn down to fibre. After Argyros et al. [2005a].

stacked structure when drawn to fibre, resulting in significant structural distortion. This problem was subsequently avoided by annealing the capillaries to alleviate any stress and a six-layer hexagonally packed structure with a high degree of regularity was produced [Huang et al. 2004a,b]. Single-mode mPOF has been drawn from stacked polystyrene preforms [Shin et al. 2004], and the technique has also been used with perfluorinated polymer [Kondo et al. 2004]. In the latter case, five-layer hexagonal solid-core mPOFs were drawn. A high air fraction was achieved through preform pressurisation.

More recently the first stacked capillary hollow-core mPOF was fabricated (see also Chapter 8). An extruded PMMA tube of 80 mm diameter was stacked with 4 mm diameter extruded capillaries with a spacer in the core. The preform is shown in Fig. 5.4. This was drawn to a fibre that showed a kagome lattice as observed in glass hollow-core PCF. Bandgap guidance was observed from 1200 to 1500 nm for these hollow-core mPOFs.

5.1.3 Drilling Monolithic Preforms

Drilling is a straightforward approach to creating a structured preform. Indeed the world's first photonic bandgap crystal was fabricated by drilling out a block of high refractive index dielectric material [Yablonovitch et al. 1991]. Glass however is a brittle material and tends to fracture if even small cracks form in the radial direction during the drilling process. Drilling has been used for a variety of glasses, including soft glasses [Tajima et al. 2003], although the success of capillary-stacking techniques has seen drilling all but vanish as a means of fabricating silica microstructured preforms [Feng et al. 2005].

For polymers, drilling methods have been very successfully used by several groups [Large et al. 2001, Barton et al. 2004, Jensen et al. 2005] for a range of materials. Figure 5.5 shows an example of drilling a preform in PMMA.

Although not a viable route for commercial mPOF production, a diversity of complex preforms can be accurately made in an automated fashion using drilling with a computer numerical control (CNC) mill [Barton et al. 2004].

Fig. 5.4. 80 mm diameter stacked-capillary mPOF preform with a close-packed hexagonal hole structure and a hollow core. Note that the (larger) central capillary is a spacer that is only present at the ends of the preform. The smaller preform on the right is a typical glass capillary stacked preform.

Fig. 5.5. CNC drilling of an mPOF preform with a triangular hole structure (left) and two examples of completed preforms (right).

The use of coated bits allows quite deep holes to be produced with minimal "drill wander" while leaving the inside of the holes with an acceptably smooth finish, the latter being of importance in that it minimises the likelihood of surface roughness induced scattering in the drawn fibre [Barton et al. 2003]. Hole sizes at the preform stage are typically 1-10 mm in diameter. Currently, the finest primary preform structure that can be drilled involves 1 mm holes with an 0.1 mm wall thickness between holes to a depth of 65 mm. The longest preform that has been drilled is 140 mm in length with a hole spacing of 2.5 mm using 2 mm drills with a 70 mm flute. This preform length was achieved by drilling from both ends of the preform.

Figure 5.6(a) shows the design of a Graded-Index mPOF (or GImPOF) as programmed into the CNC mill. Such designs can require several hundred closely spaced holes of different sizes to be drilled - a process that can take several days to complete. Figure 5.6(b) shows an optical micrograph of the drawn fibre. It is worth noting the extent and nature of the hole deformation within the final microstructure - an issue that will be dealt with in more detail in the following chapter.

Drilling monolithic preforms allows for the rapid prototyping of new mPOF designs, making it an ideal research technique. This approach can readily produce structures that are difficult if not impossible to produce by capillary stacking. However drilling preforms does have a number of serious limitations, the most obvious being that the process can be slow while the preforms produced are unusually short and fat, typically having dimensions of 70 mm for both their diameter and length. One consequence here is that to achieve the desired hole or core dimensions generally requires a two-stage drawing process where the intermediate "cane" from the first stage is sleeved within a tube before being drawn a second time to fibre [Zagari et al. 2004b, Barton et al. 2004]. This two-stage drawing process is discussed in Section. 5.2.

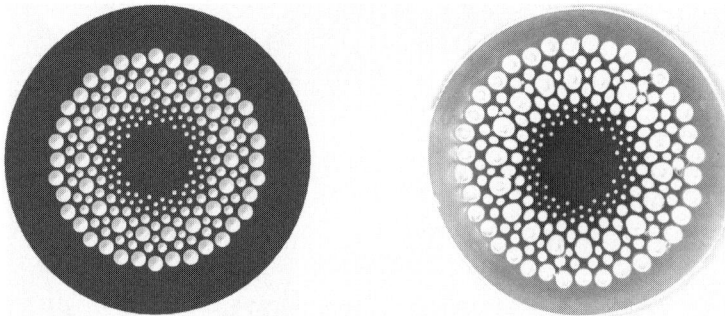

Fig. 5.6. (a) Design of an 80 mm diameter GImPOF preform. (b) 220 μm diameter GImPOF fabricated from such a preform. After Barton et al. [2004].

A less obvious limitation is that the drilling process can have a major impact on the optical performance of the final fibre [Barton et al. 2003]. This point is discussed further in Chapter 7 with Fig. 7.10 showing the transmission loss measured for a series of GImPOFs identical in terms of their drilled preform structure and final fibre dimensions but fabricated under different conditions.

Despite their research successes, neither capillary stacking nor monolithic drilling can be readily scaled up for commercial mPOF production. This task is expected to fall to technologies built around either extrusion or casting.

5.1.4 Extrusion

Recently a variety of alternatives to drilling and stacking have been considered for producing mPOF preforms with a view to fibre manufacture in commercial quantities. These include extrusion, injection moulding and casting, each of which results in a monolithic preform in which the entire hole structure is created in a single stage process.

Extrusion has been used extensively to produce structured preforms in non-silica glass materials [Allan et al. 2001, Kumar et al. 2002, Kiang et al. 2002]. Heated bulk material is forced under high pressure through a die containing a pattern consistent with the required preform structure. An alternative approach is to employ reactive extrusion. Each method is an attractive option for mPOF preform manufacture, as conceptually each should allow arbitrary hole structures to be imposed on preforms of some considerable length as necessary for commercial scale operations.

Billet extrusion

Billet extrusion has been used to fabricate microstructured polymer optical fibre preforms [Ebendorff-Heidepriem et al. 2007]. A ram extruder was used to force a polymer billet through a die to form a preform with a complex transverse profile. One advantage of this method is that it involves a single automated step by which non-circular holes, large air fractions and long preforms can be obtained. The method was demonstrated via the fabrication of a small core, high air fraction mPOF which could not be made easily by either drilling or stacking. A computer-controlled ram was used to extrude the structured preform from a 3 cm diameter commercially available PMMA billet. The billet was heated (to a temperature in the range 135 to 175 °C) and then forced (at a fixed speed in the range 0.1 to 0.5 mm/min) through a die whose geometry determined the structure of the extrudate making up the preform.

The useful temperature range for PMMA extrusion is largely limited by die swell at lower temperatures (where the extrudate diameter is larger then the die exit diameter) and by thermal degradation at higher temperatures. Die swell results in the distortion of the preform transverse profile due to elastic recovery of the polymer melt [Gupta 2000]. For a suspended core preform, die swell leads to hole and strut distortion and an increased core size relative to the preform diameter as shown in Fig. 5.7.

Although increasing the temperature and decreasing ram speed reduces die swell, in practice, higher extrusion temperatures can lead to thermal degradation, and very low ram speeds are impractical. In addition, temperature affects the surface roughness of the preform. Smooth shiny surfaces are obtained above 170 °C, whereas a matt (or 'sharkskin') surface is observed below this temperature. The latter has been attributed to flow instabilities as a result of wall slip [Gupta 2000]. At temperatures above 170 °C, the smooth

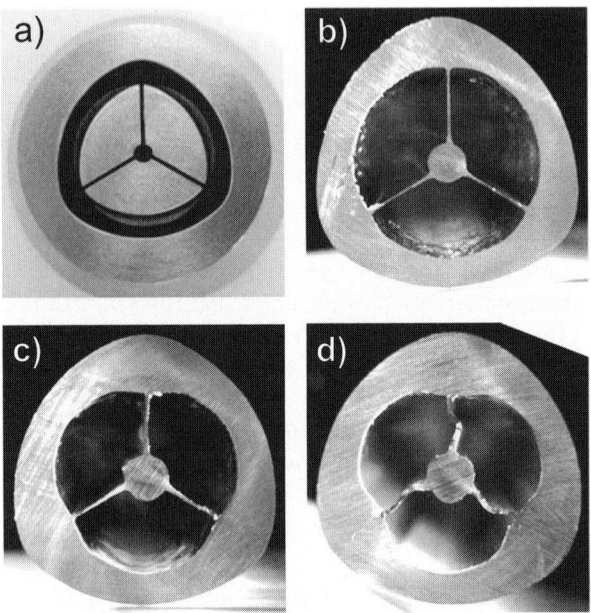

Fig. 5.7. Deviation of the preform profile from the (a) die exit profile when extruded at (b) 175 °C, (c) 170 °C, (d) 165 °C. *Images courtesy of H. Ebendorff-Heidepriem and T Monro of the Centre of Expertise in Photonics, The University of Adelaide, Australia [Ebendorff-Heidepriem et al. 2007]*

surfaces are offset by tapering of the preform as a whole due to stretching of the softened material as it emerges from the die and experiences the weight of already extruded material. Thus, for any given die design, the choice of temperature will always be a compromise.

Reactive extrusion

A process for the production of conventional POF made from poly(butyl methacrylate) by reactive extrusion has recently been developed [Berthet et al. 2006]. Reactive extrusion involves chemical reactions whereby the feed is modified either by changing its molecular weight (by polymerisation) or by combination of several polymers. The polymerisation approach has been used to fabricate large PMMA mPOF preforms [Wang et al. 2005b]. A viscous "prepolymer" was created from MMA monomer and benzoyl peroxide (an initiator) at 85-90 °C over a 6 hour period. This was then put into a reactive extruder tower and gradually heated from 80 to 185 °C until complete polymerisation was achieved after about 12 hours. The final stage involved pressing the highly viscous PMMA into a pre-warmed mould (see Section 5.1.5). Using this approach, mPOF preforms could essentially be created on a continuous basis simply by exchanging each mould when filled. The hole structure (comprising 36 holes each 2.5 mm in diameter) realised for these reactively extruded preforms was of an acceptable quality and the overall process is highly efficient with one 25 kg batch of MMA producing 14 preforms each 350 mm in length with an outer diameter of 70 mm. As this work is at an early stage, there are few experimental results from fibres drawn from these preforms.

5.1.5 Injection Moulding

Injection moulding uses the heat and pressure generated by a screw to inject molten polymer into a mould. It is one of the most commonly used processes for producing plastic goods because of the quality of the surfaces obtained and the fact that it can be used for very complex mouldings [Goodship and Love 2002].

An injection moulder consists of a heated barrel fitted with a rotating screw which feeds the molten polymer into a temperature controlled split mould. Thus the screw both plasticises the polymer and acts as a ram during the injection phase. The injection pressure required is high, up to 10^2 MPa depending on the material being processed. The mould is warmed before the injection phase with the plastic being injected quickly to prevent it hardening before the mould is full.

A given mould may be costly to design and produce, although for large production runs this is clearly a highly economical route. The mould is typically made from stainless steel, constructed so as to allow it to be readily opened and the moulded item easily removed. The internal cavity must be smooth and highly polished to ensure good surface quality for the moulded item which is usually removed using ejector pins built into the mould. In the case of preform fabrication, a set of structure generating elements (e.g. wires, rods or tubes) within the mould cavity are used to create the required hole pattern. These elements may be designed to allow individual removal once the preform polymer has cooled and solidified. Alternative removal options include the use of conically shaped elements, coating the elements with a low-adhesion material such as Teflon, heating the elements by running a current through them, or using cooling to exploit thermal expansion differences to loosen them.

Injection moulding has been used to fabricate an mPOF preform [Wang and Chen 2005]. The mould was connected directly to a single-screw extruder having a 70 mm diameter exit nozzle. Before use, the PMMA feed pellets were dried for 15 hours at 70 °C. The optimal conditions (as measured at the exit nozzle) were found to be a temperature of 185 °C and a pressure of 18 MPa. The PMMA was fed into the pre-heated mould (also at 185 °C), creating a preform 70 mm in diameter, 400 mm in length and having a total of 36 holes. A slight yellowing was observed in the final preform, possibly brought about by heat-induced oxidisation. However the hole structure in the preform was of higher quality than that obtained using in-situ polymerization of monomer in the same mould using casting (see Section 5.1.6). It was felt that this technique could be readily scaled up for commercial preform fabrication, and work is underway on large preforms containing more complex hole patterns.

5.1.6 Casting

Casting is an option that has been used to produce both glass and polymer preforms. This approach offers advantages relative to bundling or stacking methods since the hole pattern, size and spacing can be altered independently while it does not create any interstitial holes within the lattice. It also allows the best prospects of maintaining material purity, as a closed reaction vessel can be kept relatively free of contaminants such as dust.

For glasses, preform casting involves the use of (low temperature) sol-gel technology [Bise et al. 2002, Arimondi and Roba 2005, Windeler 2005, Bise and Trevor 2005].

For polymers, the necessary chemical precursors (i.e. monomer, initiator and chain-transfer agent) are introduced into a mould to produce the required hole structure. The polymerising mixture in the mould generally requires degassing to avoid bubble formation. After polymerisation is complete, the solid structure is removed from the mould. The casting of mPOF preforms was demonstrated very early on [Choi et al. 2001] with fibre being drawn from a PMMA preform (50 mm in diameter and 250 mm long) containing a hexagonal pattern of four rings of holes. More recently a large effective area mPOF was fabricated through casting of a structured preform containing 30 holes in a two-layer hexagonal arrangement [Asnaghi et al. 2002], the fabrication details of which appeared in a recent patent publication [Arimondi et al. 2005].

Many variations exist, such as casting around a structure that will remain part of the preform (thus eliminating the need to remove the hole forming elements). Examples here include casting around a stack of capillaries to fill in the interstitial holes [Large et al. 2001], or casting into the gap between a structured cane and a sleeve to create a seamless yet drawable join between the two. However with any such combined casting process, care must be taken not to destroy the structured polymer by exposure to the monomer (which is generally a solvent for the polymer). This latter problem can be avoided by using a monomer saturated with the polymer, as has been demonstrated for PMMA [Zagari 2003].

MPOF preforms have also recently been produced by casting in China [Wang et al. 2005a, Ren et al. 2005, Zhang et al. 2005, 2006]. Figure 5.8) shows a 40 mm diameter mould for fabrication of 18 and 36 hole preforms and a 75 mm diameter mould for the fabrication of a larger 60 hole preform. These moulds employ a glass outer tube (alternatively Teflon or steel can be used) with an overall length of 35 to 50 cm. Metal rods (as small as 1.5 mm in diameter) to create the preform hole structure are fixed to upper and lower Teflon cover plates attached to the mould.

Preform production by casting does however require care to be taken in material preparation and handling so as to minimise contamination. The group of Wang and co-workers used a combination of distillation and filtration to purify the monomer MMA, the polymerisation initiator benzoyl peroxide (BPO) and the chain transfer agent n-dodecyl mercaptan (DDM). The MMA

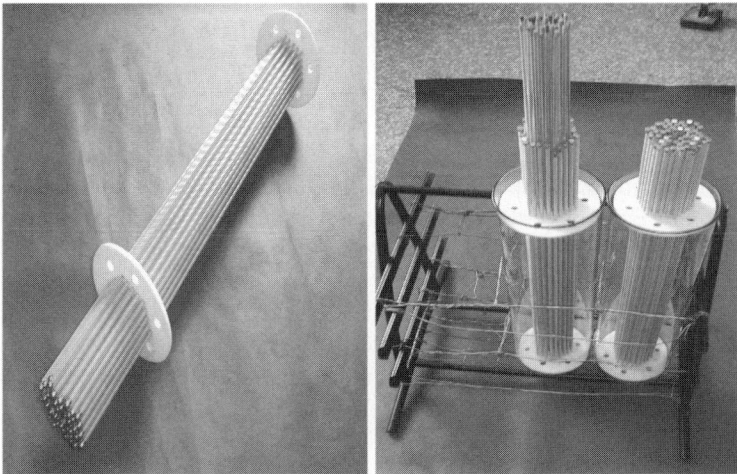

Fig. 5.8. Moulds for casting of structured polymer preforms. *Images courtesy of Dr. Lili Wang, Xi'an Institute of Optics and Precision Mechanics [Zhang et al. 2006].*

was also "rinsed" with aluminium oxide to reduce the amount of water (or other possible polar material) present. After material purification, the MMA was first converted into a pre-polymer of relatively low molecular weight by heating a MMA-BPO-DDM mixture at 90 °C for 4 hours. The pre-polymer was then cast into the mould where the polymerization process was completed. To prevent bubbles from becoming trapped within the polymerising material, gentle shaking was used in the oven. The final post-polymerisation stage involved holding the filled mould at 90 °C for 48 hours. A typical cast preform required about 500 ml of MMA, 1.25 g of BPO and 12.5 ml of DDM.

Large diameter mPOF preforms can clearly be made using casting techniques. Some technical issues remain, such as preventing the inclusion of gas pockets/bubbles, ensuring uniformity of polymer properties throughout the preform and ensuring good surface quality within the hole structure. In addition, some monomer may remain "dissolved" in the preform due to incomplete polymerisation or to depolymerisation. The presence of monomers can cause bubble formation during the fibre draw. Monomer-induced bubbling can be a major problem during fibre drawing even for drilled preforms in cases where a low-quality PMMA is used (see Table 5.2). The eventual route for manufacturing the large preforms needed for commercial mPOF production is still an open technical/economic question. However despite several unresolved issues, casting has the potential to become a realistic commercial option.

5.2 Fibre Drawing And Furnace Design

In the fibre draw, a large-scale structured preform is reduced to fibre dimensions. This process is virtually independent of the way the preform was fabricated, while a change in the material used leads (to first order) only to a difference in draw temperature. The preform and fibre fabrication stages can therefore be considered as essentially decoupled. This section will focus on techniques for drawing structured preforms to fibre using examples from drilled PMMA preforms.

The mPOF draw process is essentially the same as that used for conventional fibres. The preform is fed into a furnace at a rate that allows the temperature of polymer to be raised sufficiently above its glass transition temperature (T_g) to allow its viscosity to fall to a level where the polymer can be readily deformed. As the heated material is pulled down, a characteristic neck-down region forms where a rapid change in diameter occurs (see the example shown in Fig. 5.9). The length of this neck-down region is set by a number of parameters including the dimensions and heating intensity of the furnace, the rate at which material is passing through the furnace, and the thermal properties of the polymer. Of particular importance is the temperature dependence of the polymer's viscosity. The shape of the deforming preform within the neck-down region is critically important when drawing microstructured fibres because of its impact on hole deformation. A key consideration when seeking to minimise deformation of the microstructure during the drawing process is the need for a radial temperature profile that is as uniform as possible. The presence of a large air fraction within an mPOF preform may reduce heat transfer in the radial direction due to the low thermal conductivity of the air within the microstructure [Lyytikäinen et al. 2004]. The furnace design and operation are therefore key considerations for any mPOF draw tower and will be a primary focus for the rest of this chapter.

Drilled preforms are typically "short and fat" with a length to diameter ratio of around unity. Here they are referred to as a primary preform. The relative diameters of the primary preform and the final fibre poses challenges, and a two or three stage draw process is sometimes used. Figure 5.10 shows a schematic of the various options for drawing these preforms to fibre - with the alternatives being [1] a direct draw, [2] a stretch and draw, and [3] a stretch, sleeve and draw.

It is possible to draw a primary preform straight to fibre (route [1] in Fig. 5.10). However an intermediate step is usually added in which the preform is stretched to create a secondary preform (or "cane") of 6 to 12 mm diameter before drawing to fibre (route [2] in Fig. 5.10). Drawing to fibre is fundamentally the same process irrespective of whether a primary or a secondary preform is used. However a direct draw to fibre from a primary preform sees the diameter reduced by about 500× in a single step, requiring a factor of 250×10^3 between the draw and feed speeds [cf. Eq.(5.1)]. By contrast, drawing from a secondary preform only reduces the diameter 80× with the draw

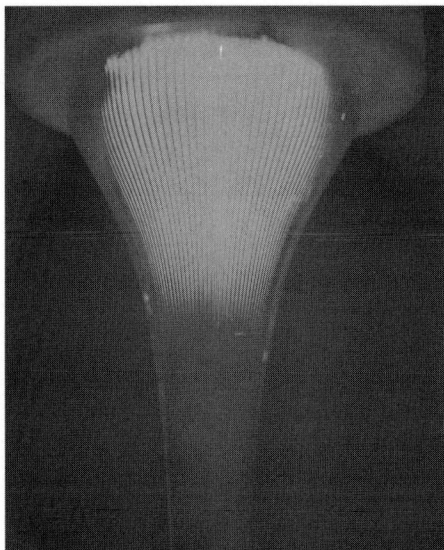

Fig. 5.9. An mPOF preform neck-down region.

and feed speeds differing by a factor of 6400×, allowing much more control over the draw process.

Many mPOF designs (e.g. the small-core fibre described in [Zagari et al. 2004b]) require that the final hole sizes be of the order of a micron (or less), and hence a sleeving stage is added to the fabrication process (route [3] in Fig. 5.10). Here a secondary preform of 2 to 6 mm diameter is drawn from the primary preform. This is referred to as a microstructured cane. This cane is sleeved within a close fitting PMMA tube to form a 12 mm secondary preform that is then drawn to fibre.

Figure 5.11 is a schematic diagram of a generic mPOF draw tower. Many variations of this generic scheme are possible, and these variations may profoundly affect the quality and extent of hole deformation in the drawn fibre. For this reason we compare three different systems, particularly with the aim of revealing the role of the furnace design. This is the major difference between three systems discussed here, with two using convective heating of the preform, the other radiative heating.

5.2.1 Convective Heat Transfer

Under steady-state operating conditions, mass conservation dictates the following relationship between the preform feed rate v_{preform}, preform diameter D, fibre draw speed v_{draw} and fibre diameter d:

$$\frac{D^2}{d^2} = \frac{v_{\text{draw}}}{v_{\text{preform}}} \tag{5.1}$$

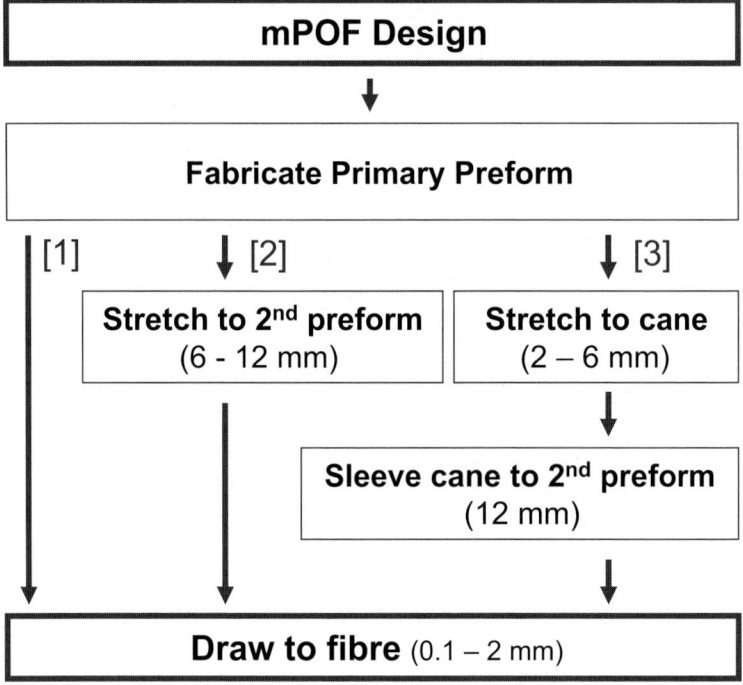

Fig. 5.10. Options for drawing an mPOF preform to fibre. After Barton et al. [2004].

Thus manual control of the fibre diameter should be achievable simply by accurately setting the feed and draw rates for a set preform diameter. In practice however, the critical parameter is the length of the hot-zone, which is defined as that region in the furnace that is sufficiently above the glass transition temperature. Based on experience, a threshold temperature of 145 °C is an appropriate choice.

The first and simplest drawing system used a heat-curing oven (i.e. to cure the polymer coating on silica fibres) as its furnace [Zagari et al. 2004b]. The furnace size, the lack of a separate pre-heater for the entering preform, and the low intensity convective heating together resulted in a long hot-zone. For this furnace, Fig. 5.12 shows the temperature profile measured using thermocouples embedded in a stationary preform for four nominal furnace temperatures. The horizontal axis indicates the distance from the top of the furnace where the preform enters. At a nominal furnace temperature of 180 °C, the hot-zone was approximately 140 mm in length. Such a long hot-zone can lead to operational problems due to movement of the preform neck-down region within the hot-zone. This movement impacts on the material balance over the furnace and thus on the observed fibre diameter. The first two rows in Table 5.1 show experimental data taken from this draw system. These clearly indicate that

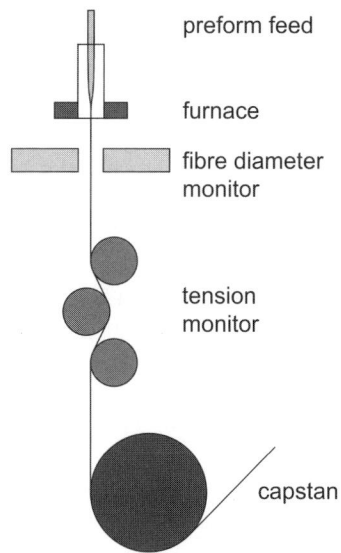

preform feed

furnace

fibre diameter
monitor

tension
monitor

capstan

Fig. 5.11. Schematic diagram of an mPOF draw tower.

even after a considerable length of time, steady-state operating conditions (in terms of mass conservation and the applicability of Eq. (5.1)) had not been achieved. Such behaviour can be explained by the slow accumulation (or depletion) of material over time within the furnace's hot-zone. This in turn leads to a slow dynamic response in terms of the observed fibre diameter.

Table 5.1. Comparison of fibre diameters for two different draw systems with values calculated using the mass conservation Eq. 5.1. Both furnaces employed convective heating of the preform.

Preform ⌀ (mm)	Feed rate (mm/min)	Draw speed (m/min)	Meas. fibre ⌀ (μm)	Calc. fibre ⌀ (μm)
12.5	10	2	1130	884
12.5	3	3	245	395
8.0	3	4.8	202	200

An example of a furnace specifically designed to have a very short hot-zone is schematically illustrated in Figure 5.13. The furnace has separate preheat and drawing sections, separated by an adjustable iris. Additional irises isolate each of these sections from the ambient atmosphere. A "tight" hot-zone was created within the drawing section using impingement heating with heated nitrogen being directed onto the preform through a ring of closely spaced

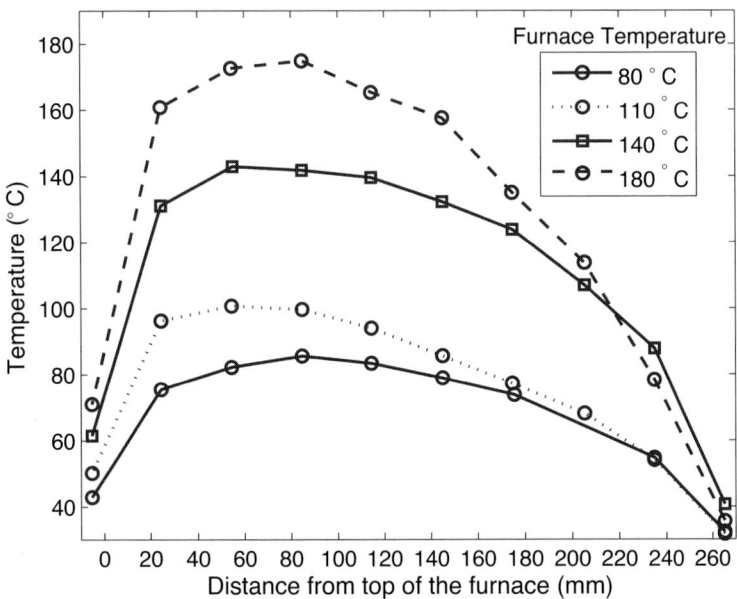

Fig. 5.12. Temperature profiles measured in a drawing furnace with a 'long' hot-zone. After Barton et al. [2004].

holes. Although air would be a cheaper option, the use of inert nitrogen was expected to minimise any surface oxidation of the PMMA as it passed through the furnace. In this design, the nitrogen is pressure regulated and split with separate lines going to the preheat and drawing sections. Each gas flow is monitored and controlled before passing through a separate insulated heating system built around a 500 W cartridge heater.

Figure 5.14 shows two temperature profiles measured within this furnace by inserting a K-type thermocouple into a 12 mm diameter PMMA rod and drawing this down to 800 µm fibre. The temperature of the heated nitrogen (entering at 205 and 231 °C) needs to be substantially above that of the deforming preform as it passes through the furnace, as forced convection is the dominant mode of heat transfer between the two. The external heat transfer coefficient (which dictates the rate of heat transfer from the nitrogen to the preform surface) is determined by the local fluid velocity within the furnace. In Fig. 5.14 the horizontal axis shows the distance from the top iris which nominally separates the preheat and drawing sections. The preheat section raised the preform temperature to around 105-110 °C, a range chosen to ensure no preform deformation prior to entering the draw section. With a separate preheat section and higher intensity convective heating, the length of the hot-zone (where the PMMA is above 145 °C) decreased to around 15-20 mm, compared to 140 mm for the furnace described previously.

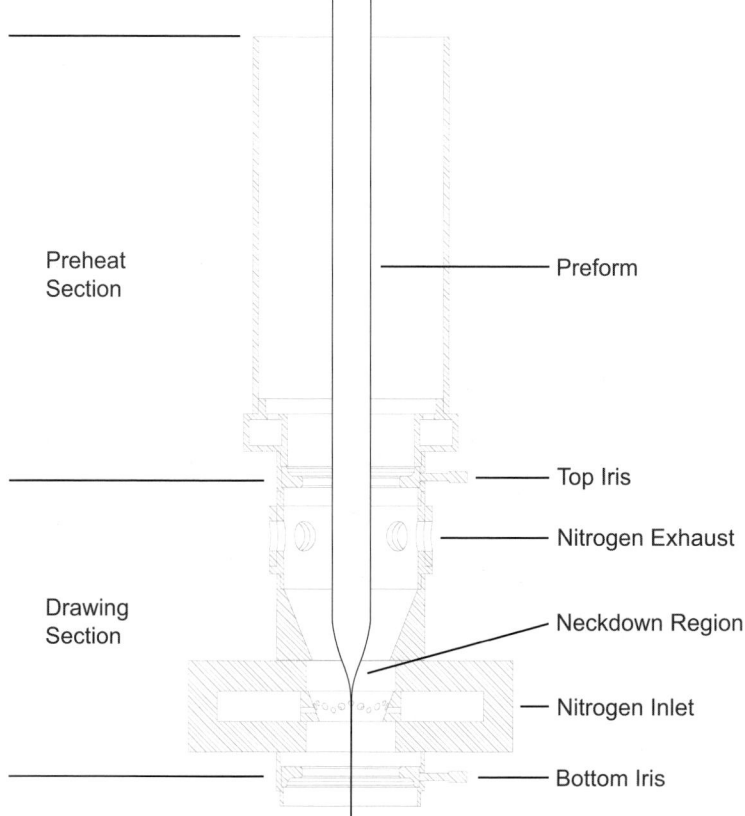

Fig. 5.13. Schematic diagram of a furnace employing convective heating in separate preheat and drawing sections. After Barton et al. [2004].

A major consequence of having a tighter hot-zone is significantly steadier operation of the drawing process with the position of the neck-down region being essentially fixed and little in the way of material accumulation or depletion within the entire drawing section. Indeed as shown in the last row of Table 5.1, the measured and calculated fibre diameters are in good agreement, even without feedback control.

This purpose-built drawing system employing a convective heating furnace has been successfully used to draw a wide range of mPOFs [van Eijkelenborg et al. 2003, Zagari et al. 2004b].

5.2.2 Radiative Heat Transfer

An alternative approach to using a preheat section is to improve the heat transfer by using radiation rather than convection. Fig. 5.15 shows a photograph a mPOF draw tower using this approach. The large size of the mPOF

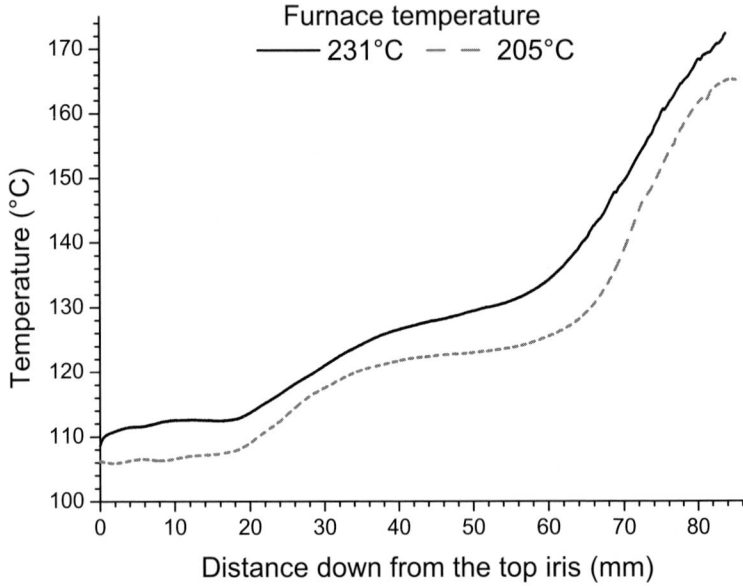

Fig. 5.14. Temperature profiles measured in a drawing furnace with a 'short' hot-zone. After Barton et al. [2004].

preforms means that drawing is still primarily done using a two-stage process. This requires the use of a large bore furnace for the preform to cane stretching, while a smaller bore furnace is used to draw cane to fibre. In each furnace, heating is provided by a series of radially located halogen lamps arranged on the outside of a quartz tube so as to provide a nearly uniform heat flux profile around the preform. The length of the hot-zone can be adjusted by the installation of an opaque shield around the quartz tube. Like the use of impingement convective heating, such an arrangement also provides a tight hot-zone within which the preform is necked down to fibre. This radiative draw tower has essentially the same functionality as the previous convective system except that it also allowed internal pressurisation of the cane during drawing. As will be discussed further in the next chapter, this modification can be used to great advantage in cases where the optical performance of the final fibre is highly dependent on the thickness of the material bridges between adjacent holes.

The use of radiative heat transfer provides two major advantages. The first is that no separate preheat section is required. This is important as considerable expertise is required with a forced convection heating system to adjust the nitrogen flow, temperature and distribution (between the preheat and drawing sections) to ensure good drawing conditions. The second advantage relates to the speed with which heat can be transferred into the interior of the preform or cane. Most of the thermal radiation from the halogen lamps is at

Fig. 5.15. Polymer draw tower with a radiative furnace.

wavelengths for which PMMA is opaque and thus this heat flux merely heats up the external surface in a similar manner to that provided by convective heating. However there are enough windows in the PMMA absorption spectrum to allow a useful amount of thermal radiation to be absorbed beyond the surface [Lwin et al. 2006]. This enhanced rate of internal heat transfer not only brings the inner portion of the cane up to the drawing temperature more quickly but it promotes a more uniform radial temperature (i.e. a smaller radial temperature gradient) which is a key objective from a fabrication perspective where the aim is to minimise deformation of the drawn fibre's microstructure.

As will be discussed in the following chapter, this radiative draw tower has been used to fabricate mPOFs that cannot be made using the earlier convective systems.

5.2.3 Monitoring And Control

Fig. 5.11 is a schematic representation applicable to all the draw systems described here. The specifics and results in the following discussion relate to the draw tower using the convective furnace shown in Fig. 5.13.

The centred preform is driven into the draw system at a feed rate (typically in the range 2-10 mm/min) set by a servo motor. The heated nitrogen supplied to the preheat and drawing sections is both flow and temperature controlled. As it is not possible to obtain online measurements of either the temperature of the preform as it leaves the preheat section or the polymer temperature as the preform necks down, the flow and inlet temperature of the nitrogen to both these sections is based on experience.

The diameter of the drawn fibre is readily measured online using a laser-based system, such as an *Anritsu* KL151A. Such a system nominally takes measurements at a frequency of 1000 Hz, although the default option of employing simple signal averaging over sets of 128 data points (giving an effective sampling rate of 7.8 Hz) is generally more than adequate for process monitoring and feedback control purposes.

After the diameter measurement, the fibre is passed through a tension monitoring system. A three-wheel *Check Line* monitor with a 0-500 g operating range is adequate. There is a trade-off between draw temperature and draw tension with the former being adjusted to maintain tension in the range 5-200 g. The difference in the measured peak temperatures shown in Fig. 5.14 is small. However this modest temperature change results in less than half the applied tension being needed to draw the fibre (145 g for 165 °C and 65 g for 172 °C, respectively). As will be discussed in the next chapter, the temperature of the polymer as it is drawn from preform to fibre, along with the necessary drawing tension, together play a critical role in determining the extent and form of any mPOF structural deformation.

The most commonly used control option when drawing an mPOF is to maintain fibre diameter using a (two-term or proportional-integral) feedback controller that adjusts capstan speed (which is equivalent to fibre draw rate) on the basis of the calculated deviation between the measured and set-point diameter values. In this control configuration, the fibre tension level can be changed by manually adjusting the set-point value on the feedback controller maintaining the temperature of the nitrogen flowing into the drawing section of the furnace. An alternative configuration whereby the fibre tension (rather than the fibre diameter) is controlled by adjusting the capstan speed can also be used, though this leads to a less uniform fibre diameter. There are a number of commercial software packages that can be used to implement this process monitoring and control. *LabView* was used for the mPOF draw towers described here.

5.3 Quality Of Fabricated MPOF

5.3.1 Material

Although tight process monitoring and control are necessary when drawing a preform to fibre, successful mPOF fabrication is also intimately linked with material selection. Whatever the polymer used in fabricating an mPOF, material quality will impact on fibre performance. Considerable experience in this regard is available for mPOF fabricated from PMMA and is outlined here.

In order for a thermoplastic to be drawable, the chain length should be sufficiently low and no cross-linking should be present. For PMMA the molecular weight M_w should be around 60×10^3. This corresponds to chain lengths of around 600 monomer units. Most commercially available cast PMMA is of a very high molecular weight ($M_w \approx 10^6$) and is therefore not drawable.

On the other hand, low-cost PMMA extruded rods are easy to handle at the primary preform stage (e.g. drilling the required microstructure), but their bulk material loss is high (see Table 5.2 for representative figures). This material also often bubbles when drawn. The extent of bubbling is determined by the type of polymerisation used. In unassisted free-radical polymerisation the termination step is predominantly by disproportionation. This process leaves two neutral, unreactive polymer chains, one with a saturated end-group (i.e. single bonds), the other with a terminal double bond. The latter unsaturated chain has a greater tendency to depolymerise when heated, leading to the formation of monomer that can vaporise during the draw process. The measured glass transition temperature (T_g) such PMMA is relatively low (approximately 95 °C), consistent with the plasticising effect of monomer.

Higher quality PMMA extruded rods give very few (if any) bubbles when a preform was drawn to fibre, although its bulk material loss is still quite high (see Table 5.2). Such a difference in behaviour can be explained by this polymerisation having been carried out in the presence of a chain transfer agent (i.e. an assisted polymerisation), an approach that leads to the ends of all PMMA chains being saturated (i.e. having single bonds). These ends are highly stable and would not be expected to depolymerise when heated up during the draw process. The T_g for this material was measured as 115 °C, a value consistent with the absence of any significant amount of monomer to act as a plasticiser.

The lowest transmission loss was obtained using high purity cast PMMA rods, though this material is both harder and more brittle than the other options.

It should also be noted that the hole structure in the primary preform needs to be thoroughly cleaned out by extensive water flushing (more aggressive solvents should not be used) to remove residual cutting fluid and PMMA micro-swarf after the drilling process. Such residues have been identified as a likely contributor to measured scattering losses in the final fibre [van Eijkelenborg et al. 2004].

Table 5.2. Measured bulk material loss for three types of PMMA rod. Loss is quoted at a wavelength of 633 nm.

PMMA	Loss (dB/m)
Low-cost extruded	4.0
High-quality extruded	2.5
High-purity cast	0.67

5.3.2 Structure

The remainder of this chapter is concerned with two key mechanical criteria, fibre diameter and the extent of hole deformation, both of which impact strongly on the optical performance.

Using as an example the high intensity convectively heated furnace described previously, two operational issues emerged that were crucial to the maintenance of tight fibre diameter control. Firstly, the nitrogen flow through the drawing section of the furnace must be low enough to avoid highly turbulent flow patterns that cause the developing fibre to vibrate and/or be exposed to time dependent heat transfer. However once again a compromise solution is required as impingement heat transfer (to the preform in the hot-zone) is enhanced by high gas velocities. For our system, the optimum nitrogen flow to the drawing section was found to be 2.7 L/min. Secondly, diameter variations increased noticeably if the preform was not accurately centred as it was fed into the furnace.

Figure 5.16 shows recorded draw parameters (i.e. fibre tension, draw speed and diameter) over 37 m of a GImPOF drawn under well controlled conditions. The fibre structure was as shown in Fig. 5.6. The average fibre diameter over this length was 200 μm with a standard deviation of ± 1.7 μm. This example was chosen to illustrate the variation in capstan (or draw) speed arising from the feedback control loop responding to measured variations in fibre diameter. For a well controlled draw tower, a standard deviation of less than ± 1 μm in fibre diameter is readily achievable.

The second mechanical quality criterion considered here is the extent of hole collapse between the primary preform and the final fibre. For a range of mPOFs, Table 5.3 gives the hole size h and hole spacing Λ for the primary preform, as well as the fibre diameter and the percentage hole collapse in the final drawn fibre, the latter being defined as follows:

$$C_{\text{fibre}} = 100 \times \left(1 - \frac{(h/\Lambda)_{\text{fibre}}}{(h/\Lambda)_{\text{preform}}} \right). \tag{5.2}$$

Average hole size and hole spacing for all drawn fibres (except the small-core fibre) were based on optical micrographs. In the small-core fibre case, more accurate scanning electron microscope measurements of the hole size

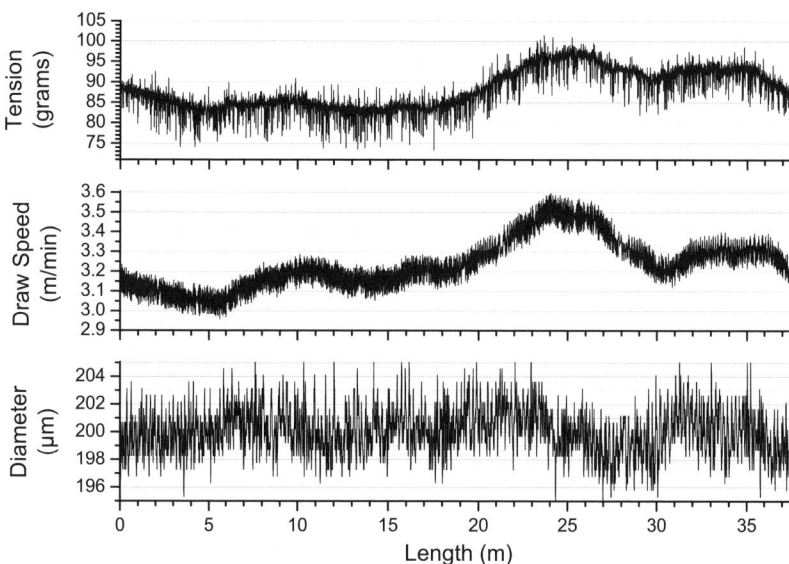

Fig. 5.16. Draw tension, draw speed and fibre diameter as recorded during a GIm-POF draw. After Barton et al. [2004].

(0.53 µm) and hole spacing (1.38 µm) within the fibre structure were also taken to allow numerical modelling to be carried out, showing the fibre to be endlessly single-mode [Zagari et al. 2004b].

Table 5.3. Data showing the hole size (h_{preform}) and hole spacing ($\Lambda_{\mathrm{preform}}$) in the primary preform, as well as the fibre diameter (d_{fibre}) and the percentage hole collapse in the final drawn fibre (C_{fibre}) for a range of fabricated mPOFs.

Fibre type	h_{preform} (mm)	$\Lambda_{\mathrm{preform}}$ (mm)	d_{fibre} (µm)	C_{fibre} (%)
Single-mode [van Eijkelenborg et al. 2001]	4.0	6.0	250	31
Single-mode [Zagari 2003]	1.2	1.7	250	38
Multicore [Zagari 2003]	1.2	1.7	200	39
Small-core [Zagari et al. 2004b]	1.0	1.2	570	54
Honeycomb [Zagari 2003]	1.2	1.7	230	30
Air-core bandgap [van Eijkelenborg et al. 2002]	1.0	1.2	500	37
Air-core bandgap [Zagari 2003]	2.0	2.5	160	38
Twin-core [Padden et al. 2004]	1.2	1.5	200	31

The experimental data given in Table 5.3 provide information on *gross* hole collapse and must be treated with caution for two reasons. The first is that each of these fibres was fabricated using a two-step process. Thus hole collapse was partly due to the fabrication of the secondary preform, and partly due to the drawing of the secondary preform to fibre. The second reason is that this data provides no insight into the way individual holes within the microstructure behave during the size reduction process. Hole deformation turns out to be considerably more complex than a geometric size reduction (i.e. from primary preform through to final fibre) with some additional hole collapse brought about by surface tension forces. The issue of hole deformation (both in size and shape) will be explored in the following chapter.

References

Allan, D C, West, J A, Fajardo, J C, Gallagher, M T, Kock, K W, and Borrelli, N F (2001). Photonic crystal fibers: Effective-index and band-gap guidance. In Soukoulis, C M, editor, *Photonic Crystal and Light Localisation in the 21st Century*, pages 305–20. Kluwer Academic Publishers, Dordrecht, The Netherlands.

Argyros, A, Birks, T A, Leon-Saval, S G, Cordeiro, C M B, Luan, F, and Russell, P St J (2005a). Photonic bandgap with an index step of one percent. *Optics Express*, 13(1):309–14.

Argyros, A, Birks, T A, Leon-Saval, S G, Cordeiro, C M B, and Russell, P St J (2005b). Guidance properties of low-contrast photonic bandgap fibres. *Optics Express*, 13(7):2503–11.

Arimondi, M, Macchetta, A, Asnaghi, D, and Castaldo, A (2005). Process for manufacturing a micro-structured optical fibre. US Patent 20050286847.

Arimondi, M and Roba, G S (2005). Process for manufacturing a microstructured optical fibre. US Patent 2005/0072192 A1.

Asnaghi, D, Gambirasio, A, Macchetta, A, Sarchi, D, and Tassone, F (2002). Fabrication of a large effective-area microstructured plastic optical fibre: design and transmission tests. In *Proceedings of the European Conference on Optical Communications*, volume 3, pages 632–3, Copenhagen, Denmark.

Barton, G, van Eijkelenborg, M A, Henry, G, Issa, N A, Klein, K-F, Large, M C J, Manos, S, Padden, W, Pok, W, and Poladian, L (2003). Characteristics of multimode microstructured pof performance. In *Proceedings of the International Plastic Optical Fibres conference*, volume 12, pages 81–84, Seattle, USA.

Barton, G W, van Eijkelenborg, M A, Henry, G, Large, M C J, and Zagari, J (2004). Fabrication of microstructured polymer optical fibres. *Optical Fiber Technology*, 10(4):325–35.

Berthet, R, Chalamet, Y, Taha, M, and Zerroukhi, A (2006). Optical fibers by reactive extrusion of butyl methacrylate. *Macromolecular Materials and Engineering*, 291(6):720–31.

Birks, T A, Knight, J C, and Russell, P St J (1997). Endlessly single-mode photonic crystal fiber. *Optics Letters*, 22(13):961–3.

Bise, R T and Trevor, D J (2005). Sol-gel derived microstructured fiber: Fabrication and characterization. In *Proceedings of the Optical Fiber Communication Conference*, volume 3, Anaheim, CA, USA.

Bise, R T, Windeler, R S, Kranz, K S, Kerbage, C, Eggleton, B J, and Trevor, D J (2002). Tunable photonic bandgap fiber. In *Proceedings of the Optical Fiber Communications Conference*, volume 70, pages 466–8, Optical Society of America, Washington DC, USA. OSA Trends in Optics and Photonics.

Choi, J, Kim, D Y, and Paek, U C (2001). Fabrication and properties of polymer photonic crystal fibers. In *Proceedings of the International Plastic Optical Fibres Conference*, pages 335–60, Amsterdam, Netherlands.

Ebendorff-Heidepriem, Heike, Monro, Tanya M, van Eijkelenborg, Martijn A, and Large, Maryanne C J (2007). Extruded high-NA microstructured polymer optical fibre. *Optics Communications*, 273.

Feng, X, Mairaj, A K, Hewak, D W, and Monro, T M (2005). Nonsilica glasses for holey fibers. *Journal Lightwave Technology*, 23(6):2046–54.

Gao, Y, Guo, N, Gauvreau, B, Rajabian, M, Skorobogata, O, Pone, E, Zabeida, O, Martinu, L, Dubois, C, and Skorobogatiy, M. (2006). Consecutive solvent evaporation and co-rolling techniques for polymer multilayer hollow fiber preform fabrication. *Journal of Materials Research*, 21:2246.

Goodship, V and Love, J C (2002). *Multi-Material Injection Moulding*. The University of Warwick, Warwick, UK.

Goto, M, Quema, A, Takahashi, H, Ono, S, and Sarukura, N (2004). Teflon photonic crystal fiber as terahertz waveguide. *Japanese Journal of Applied Physics*, 43:L317–9.

Gupta, R K (2000). *Polymer and composite rheology*. Marcel Dekker Inc., The Netherlands.

Han, H, Park, H, Cho, M, and Kim, J (2002). Terahertz propagation in a plastic photonic crystal fiber. *Applied Physics Letters*, 80(15):2634–6.

Huang, C, Ho, M, Cheng, C, Ma, K, Kiang, Y, Chang, H, and Yang, C C (2004a). Design, fabrication, and characterization of polymer microstructured fiber. In *Proceedings of the Photonics West Conference*, San Jose, CA, USA.

Huang, C W, Ho, M C, Yu, C P, Chang, H C, Yang, C C, Chien, H H, Ma, K J, and Zheng, Z P (2004b). Fabrication and characterization of microstructured polymer optical fibres. In *Proceedings of the Conference on Lasers and Electro Optics*, page CThX2, San Francisco, USA.

Jensen, J, Hoiby, J P, Emiliyanov, G, Bang, O, Pedersen, L, and Bjarklev, A (2005). Selective detection of antibodies in microstructured polymer optical fibers. *Optics Express*, 13(15):5883–9.

Kiang, K M, Frampton, K, Monro, T M, Moore, R, Tucknott, J, Hewak, D W, Richardson, D J, and Rutt, H N (2002). Extruded single mode non-silica glass holey optical fibres. *Electronics Letters*, 38(12):546–7.

Knight, J C, Birks, T A, and Russell, P St J (2001). "Holey" silica fibers. In Markel, V A and George, T F, editors, *Optics of Nanostructured Materials*, chapter 2, pages 39–71. Wiley, New York, USA.

Knight, J C, Birks, T A, Russell, P St J, and Atkin, D M (1996). All-silica single mode optical fiber with photonic crystal cladding. *Optics Letters*, 21(19):1547–9.

Kondo, S, Ishigure, T, and Koike, Y (2004). Fabrication of polymer photonic crystal fiber. In *Proceedings of the Micro-Optics Conference*, volume 10, pages B–7, Jena, Germany.

Kumar, V V Ravi Kanth, George, A K, Reeves, W H, Knight, J C, Russell, P St J, Omenetto, F G, and Taylor, A J (2002). Extruded soft glass photonic crystal fiber for ultrabroad supercontinuum generation. *Optics Express*, 10(25):1520–5.

Large, M C J, van Eijkelenborg, M A, Argyros, A, Zagari, J, Manos, S, Issa, N A, Bassett, I, Fleming, S, McPhedran, R C, de Sterke, C M, and Nicorovici, N A P (2001). Microstructured polymer optical fibres: a new approach to POFs. In *Proceedings of the International Plastic Optical Fibres conference*, page Post deadline paper, Amsterdam, the Netherlands.

Lwin, R, Barton, G W, Large, M C J, Poladian, L, and Xue, S (2006). Heat transfer in preforms for microstructured polymer optical fibres. In *Proceedings of the International Plastic Optical Fibres conference*, Seoul, Korea.

Lyytikäinen, K, Zagari, J, Barton, G, and Canning, J (2004). Heat transfer within a microstructured polymer optical fibre preform. *Modelling and Simulation in Materials Science and Engineering*, 12(3):S255–65.

Padden, W, Argyros, A, Manos, S, and van Eijkelenborg, M A (2004). Coupling in a twin-core microstructured polymer optical fibre. *Applied Physics Letters*, 84(10):1689–91.

Park, J H, Shin, B G, and Kim, J J (2002). Fabrication of plastic holey fibers. In *Proceedings of the International Plastic Optical Fibres conference*, volume 11, pages PD9–11, Tokyo, Japan.

Ren, L, Zhang, Y, Wang, X, Li, Y, Zhao, W, and Wang, L (2005). Fabrication and characteristics of rhodamine-doped micro-structured polymer optical fibers. In *Proceedings of the International Plastic Optical Fibres conference*, volume 14, Hong Kong, China.

Shin, B-G, Park, J-H, and Kim, J-J (2004). Plastic photonic crystal fiber fabricated by centrifugal deposition method. *Journal of Nonlinear Optical Physics and Materials*, 13(3-4):519–23.

Tajima, K, Ohashi, M, Kurokawa, K, Nakajima, K, and Yoshizawa, N (2003). Method for manufacturing optical fiber using ultrasonic drill. Patent 20030136154.

van Eijkelenborg, M A, Argyros, A, Bachmann, A, Barton, G W, Large, M C J, Henry, G, Issa, N A, Klein, K F, Poisel, H, Pok, W, Poladian, L, Manos, S, and Zagari, J (2004). Bandwidth and loss measurements of graded-index microstructured polymer optical fibre. *Electronics Letters*, 40(10):592–3.

van Eijkelenborg, M A, Argyros, A, Barton, G W, Bassett, I M, Fellew, M, Henry, G, Issa, N A, Large, M C J, Manos, S, Padden, W, Poladian, L, and Zagari, J (2003). Recent progress in microstructured polymer optical fibre fabrication and characterization. *Optical Fiber Technology*, 9(4):199–209.

van Eijkelenborg, M A, Large, M C J, Argyros, A, Bassett, I, and Zagari, J (2002). Photonic bandgap guiding in microstructured polymer optical fibres. In *Proceedings of the International Quantum Electronics Conference*, page QThH3, Moscow, Russia.

van Eijkelenborg, M A, Large, M C J, Argyros, A, Zagari, J, Manos, S, Issa, N A, Bassett, I, Fleming, S, McPhedran, R C, de Sterke, C M, and Nicorovici, N A P (2001). Microstructured polymer optical fibre. *Optics Express*, 9(7):319–27.

Wang, L and Chen, X (2005). Extrusion technique of microstructured polymer optical fiber preform. In *Proceedings of the International Plastic Optical Fibres Conference*, volume 14, Hong Kong, China.

Wang, L, Zhang, Y, Ren, L, Wang, X, Li, T, Hu, B, Zhao, W, and Li, Y (2005a). Chemical fabrication techniques for microstructured polymer optical fiber preforms. In *Proceedings of the International Plastic Optical Fibres Conference*, volume 14, Hong Kong, China.

Wang, X, Wang, L, Ren, L, Zhang, Y, Li, T, Zhao, W, and Li, Y (2005b). Design and fabrication of reacting extruder for preparation of high quality microstructured polymer optical fiber preform. In *Proceedings of the International Plastic Optical Fibres Conference*, volume 14, Hong Kong, China.

Windeler, R S (2005). Improving microstructured fibers including preform fabrication, fiber draw, splicing and postproduction dispersion tailoring. In *Proceedings of the international workshop on optical waveguide theory and numerical modelling*, volume 14, Sydney, Australia.

Yablonovitch, E, Gmitter, T J, and Leung, K M (1991). Photonic band structure: The face-centered-cubic case employing nonspherical atoms. *Physical Review Letters*, 67(17):2295–8.

Zagari, J (2003). The fabrication of microstructured polymer optical fibres. Masters dissertation, Chemical Engineering, The University of Sydney, Sydney, Australia.

Zagari, J, Argyros, A, Barton, G W, Henry, G, Large, M C J, Issa, N A, Poladian, L, and van Eijkelenborg, M A (2004a). Erratum on "small-core single-mode microstructured polymer optical fibre with large external diameter". *Optics Letters*, 29(13):1560.

Zagari, J, Argyros, A, Barton, G W, Henry, G, Large, M C J, Issa, N A, Poladian, L, and van Eijkelenborg, M A (2004b). Small-core single-mode microstructured polymer optical fibre with large external diameter. *Optics Letters*, 29(8):818–20. See also [Zagari et al. 2004a].

Zhang, Y, Li, K, Wang, L, Ren, L, Zhao, W, Miao, R, Large, M C J, and van Eijkelenborg, M A (2006). Casting preforms for microstructured polymer optical fibre fabrication. *Optics Express*, 14(12):5541–7.

Zhang, Y, Wang, L, Ren, L, Wang, X, Li, T, Zhao, W, and Li, Y (2005). Fabrication and characterization of microstructured polymer optical fiber with elliptical core. In *Proceedings of the International Plastic Optical Fibres Conference*, volume 14, Hong Kong, China.

6

Effects of Drawing on the Microstructure

"Drawing is the honesty of the art. There is no possibility of cheating. It is either good or bad."

Salvador Dali (Spanish painter, 1904-1989)

The stretch and drawing processes used to make mPOF pose considerable computational challenges, combining as they do non-isothermal, three-dimensional (3-D) and time dependent behaviour. This complexity is further complicated if the materials used exhibit significant nonlinear viscoelastic behaviour, while the very nature of an mPOF means that its fabrication involves the substantial deformation of a (potentially large) number of 3-D free surfaces. Unsurprisingly the relevant analytical [Fitt et al. 2002] and numerical [Deflandre 2002, Lyytikäinen et al. 2004] literature is limited. However a convincing story has recently emerged that quantitatively ties together the roles of material properties and tower draw conditions in determining both hole size and shape deformation within an overall MOF structure. This chapter begins with a scaling analysis leading to a suite of dimensionless numbers whose values can be used to assess the relative importance of the viscous, inertial, gravitational and surface tension forces in any particular drawing process. Subsequently both isothermal and non-isothermal drawing are considered. In the former case indicative theoretical analysis is possible, while in the latter it is necessary to rely on numerical modelling. Both hole size and shape changes are considered. Although the focus is on polymer fibres, consideration is also given to silica based microstructured optical fibres so as to highlight the impact of differing material properties. The chapter concludes by considering the impact of hole pressurisation during the draw process.

6.1 Scaling Analysis For Fibre Drawing

In general any fibre fabrication process can be considered as a transient or steady-state laminar flow of an incompressible material. The equations governing such a flow comprise three-dimensional mass, momentum and energy balances together with a description of the material's rheological behaviour. To fully specify the numerical description of the drawing process, it is also necessary to impose appropriate kinematic, dynamic and thermal boundary conditions, as well as an appropriate set of initial conditions if the case being considered is transient. The full governing system of equations has to be solved numerically if no further assumptions or simplifications are made. However the computational load can be reduced (and the numerical analysis more oriented with the reality of fibre drawing) if the less influential terms are removed from the system of equations. This relative ranking of terms within the fibre drawing model can be achieved by carrying out a "scaling analysis" on a representative (of any complex hole structure) case, the most obvious being the non-isothermal drawing of an axisymmetric annular fibre. After variable scaling (following [Schultz and Davis 1982]) and assuming that all material properties are constant (except for viscosity which is strongly temperature dependent), the dimensionless form of the equations governing fibre drawing reveals the following set of dimensionless numbers where each relates key mechanisms occurring within the overall drawing process [Xue et al. 2005a]:

- Aspect ratio $\varepsilon = R_{\text{preform}}/L$ (the "slenderness" of the extending preform or cane)
- Reynolds number $Re = \rho v_{\text{preform}} R_{\text{preform}}/\mu$ (the ratio of inertial and viscous forces)
- Froude number $Fr = v_{\text{preform}}^2/gR_{\text{preform}}$ (the ratio of inertial and gravitational forces)
- Peclet number $Pe = \rho c_p v_{\text{preform}} R_{\text{preform}}/\kappa$ (the ratio of convective to conductive heat transfer)
- Brinkman number $Br = \mu v_{\text{preform}}^2/\kappa T_{\text{ref}}$ (the ratio of heat production by mechanical dissipation to heat transfer by conduction)
- Capillary number $Ca = \mu v_{\text{preform}}/\sigma$ (the ratio of viscous and surface tension forces)
- Biot Number $Bi = hR_{\text{preform}}/\kappa$ (the ratio of internal and external resistances to heat transfer in the radial direction)

By examining the relative order of magnitude of these dimensionless numbers under typical operational conditions, it is possible to infer which forces (e.g. inertial, gravitational and surface tension) and mechanisms (e.g. external convective and internal conductive heat transfer) might be expected to play an influential role in any given fibre draw process.

For example, Table 6.1 shows conditions for the steady-state drawing of PMMA (from intermediate cane to fibre) and silica from preform to fibre. R_{preform} and R_{fibre} are the initial and final radii while T_{w} is a representative drawing temperature - taken here as that of the hot gas used to convectively heat the PMMA (see Section 5.2.1) and the wall temperature of the heating furnace for silica. L is the length of the heating zone while v_{preform} and v_{draw} are the feed and fibre draw velocities, respectively.

Table 6.1. Representative microstructured fibre drawing conditions. The mPOF draw conditions are based on the convective draw process described previously(see Section 5.2.1).

Parameter (units)	PMMA fibre	Silica fibre
R_{preform} (mm)	5	12.5
R_{fibre} (mm)	0.240	0.0625
T_{w} (°C)	220	2000
L (mm)	40–50	40–50
v_{preform} (mm/min)	2.3	3
v_{draw} (m/min)	1	120

As noted, for scaling analysis purposes only, the temperature dependence of viscosity is considered as this may vary by several orders of magnitude over the temperature range experienced during the preheating and drawing processes, with all other material properties taken as constant at an appropriate reference temperature T_{ref}. Here it is assumed that viscosity obeys an Arrhenius type dependency:

$$\mu\left(T\right) = \mu\left(T_{\text{ref}}\right)\exp\left[\frac{\Delta H}{R_g}\left(\frac{1}{T} - \frac{1}{T_{\text{ref}}}\right)\right] \tag{6.1}$$

where ΔH and R_g are the activation energy and gas constant, respectively, and T_{ref} is taken as 443 K for PMMA and the softening point of 1900 K for silica.

Thus Table 6.2 gives representative material properties adequate for scaling analysis purposes. For PMMA the surface tension coefficient measured by Wu [1970] is given with other material properties as cited by Reeve et al. [2001, 2003]. For silica the viscosity at the reference temperature is calculated from the relationship given by Bansal and Doremus [1986] while the ratio H/R_g has been determined from data given by Myers [1989]. The temperatures cited are the glass transition for PMMA and the softening point for silica.

Table 6.2. Representative material properties.

Property	PMMA	Silica	Units
Density (ρ)	1195	2200	kg/m^3
Viscosity (μ)	1.5×10^5	8.3×10^6	Pa.s
Surface tension coefficient (σ)	0.032	0.3	N/m
Thermal conductivity (κ)	0.2	2.68	W/m.K
Heat capacity (C_p)	1465	1345	J/kg.K
Activation energy ratio $(\Delta H/R_g)$	3218	6.1×10^4	K
Reference Temperature (T_{ref})	443	1900	K

Using such representative data, indicative dimensionless numbers can be calculated [Xue et al. 2005a] for the first stage (i.e. primary preform to intermediate cane) of the mPOF draw process. For this transient process, it can be concluded that any inertial force is always negligible while the gravitational force is only important in the early stages - indeed the latter is the initial driving force for the preform "flowing" and elongating. As preform deformation continues however, the gravitational force becomes less important, with the result that in practice elongation has to be aided via the application of an external draw tension. As far as surface tension is concerned, this force can become important in the later stages of the process after significant preform elongation has occurred.

A similar scaling analysis can be performed for the second stage (i.e. intermediate cane to fibre) of the mPOF draw process, as well as for the continuous drawing of silica MOF [Xue et al. 2005a]. Representative values for each dimensionless number in these continuous drawing cases lead to the following conclusions:

- Inertial and gravitational forces are negligible for both PMMA and silica drawing.
- Surface tension can become important in both cases as the hole radius decreases.
- External convective heat transfer is such that temperature gradients in the axial direction are small. This conclusion is still valid in the silica drawing case as although thermal radiation dominates, to a first approximation this heat transfer mode can be approximated as an equivalent convective flux driven by the temperature difference between the furnace wall and the surface of the silica. From a numerical modelling perspective, this result is important as it implies that a zero temperature gradient boundary condition can reasonably be imposed at the exit to the draw zone.
- In both mPOF and silica MOF, the radial temperature gradient at any axial position is expected to be small. However the fibre temperature at any point in the drawing process is expected to be sensitive to the external environment (i.e. to the temperature and flow fields around the fibre). This means that the thermal boundary conditions at the external free surface will have a critical impact on the development of the neck-down profile and thus on the extent of internal hole deformation - as will be discussed later.
- Mechanical dissipation can be safely ignored for mPOFs but not when drawing silica MOF.

Such indicative scaling analysis is invaluable as not only does it permit a significant simplification in any subsequent numerical modelling but it also provides insights into the key processes and mechanisms that together define both fibre drawing and hole structure deformation.

Based on this scaling analysis, Xue and co-workers have used both rheological analysis and modelling, in parallel with detailed experimental measurements, to study the two-stage mPOF draw process. The primary preform to intermediate cane stage was modelled as a transient draw process [Xue et al. 2005a], while a steady-state approach was used to describe the subsequent drawing of the cane to fibre [Xue et al. 2005b, 2006]. Their simulation studies were carried out using a commercial computational fluid dynamics (CFD) package [PolyFlow 2003].

6.2 Isothermal Analysis Of Hole Deformation

By its nature, the drawing of microstructured fibre is a non-isothermal process involving the deformation of a viscoelastic material containing a significant (and possibly quite large) number of holes. Analytical treatment of such cases is clearly not possible and the only means of providing quantitative results is to resort to CFD-based rheological modelling. However it is instructive initially to consider a far simpler case - that of the isothermal deformation of a single centrally located circular hole situated within a Newtonian material.

6.2.1 Hole-Size Changes

The results from such an isothermal analysis are best appreciated via the introduction of two additional dimensionless numbers. The first χ is simply the ratio of the initial hole radius R_{hole} relative to the initial radius R_{preform} of the preform:

$$\chi \equiv \frac{R_{\text{hole}}}{R_{\text{preform}}} \tag{6.2}$$

The second dimensionless number is a collapse ratio C_o that varies with axial position (z) and is defined as the ratio of the hole radius (R_{h}) at z to the radius of the preform or cane (R_{p}) at z, this ratio being itself normalised against χ:

$$C_o \equiv \frac{R_{\text{h}}(z)}{R_{\text{p}}(z)} \Big/ \chi \tag{6.3}$$

With this definition, hole closure occurs when $C_o = 0$, hole contraction occurs when $C_o < 1$, the status quo is preserved when $C_o = 1$, and hole expansion occurs when $C_o > 1$. If the slender body assumption is realistic (i.e. $\varepsilon \ll 1$), then the maximum extent of hole collapse at the end of the draw zone can be related to both the material properties and fibre drawing conditions via the following [Xue et al. 2005b],

$$C_o = 1 - \left(\frac{1}{\chi}\right)\left(\frac{1}{\ln Dr}\right)\left(\frac{1}{\varepsilon}\right)\left(\frac{1}{Ca}\right) \tag{6.4}$$

where the draw ratio Dr is defined as the ratio of draw speed v_{draw} to feed rate v_{preform}. From this relationship, it is clear that small values of χ, ε, Ca and Dr all promote hole collapse.

Hole expansion (i.e. $C_o > 1$) is counter-intuitive to many people, particularly those with experience of drawing silica MOFs. In understanding why it occurs, it is important to recognise that the extending preform or cane may not qualify as a slender body. Thus, the analysis leading to Eq. 6.4 may be invalid. In such cases, CFD simulation is the only viable approach [Xue et al. 2005b]. Figures 6.1 and 6.2 show simulation results for the collapse ratio C_o as a function of ε and Ca for two different values of χ. Clearly whether hole collapse ($C_o < 1$) or hole expansion ($C_o > 1$) occurs during the draw is determined not only by the magnitude of surface tension effects (as might have been intuitively supposed) but also by the slenderness and the initial radius ratio χ. Indeed it is apparent that C_o approaches an asymptotic value (that varies with ε and χ) as Ca increases and surface tension effects become less important (being negligible for Ca above about 200).

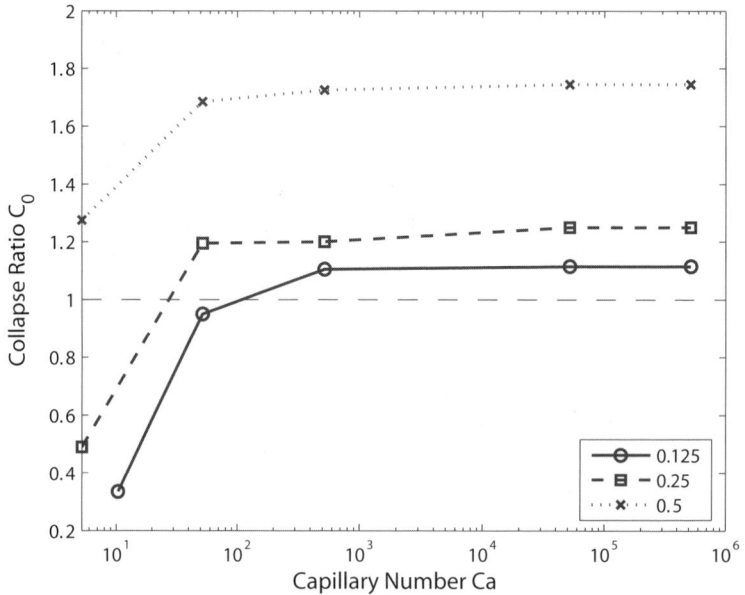

Fig. 6.1. Collapse ratio as a function of capillary number at different aspect ratios ($\varepsilon = 0.125, 0.25, 0.5$) for an initial radius ratio $\chi = 0.2$. After Xue et al. [2005b] (©[2005] IEEE).

As ε decreases, the slender body assumption becomes more reasonable and the extent of any hole expansion decreases. In the limit as $\varepsilon \to 0$ (and the slender body assumption is completely valid), so $C_o \to 1$ as $Ca \to \infty$. Both numerical simulation and experimental measurements have demonstrated that hole expansion, despite being counter-intuitive, is a very real phenomenon in cases where the slender body assumption is not reasonably met. Both also showed that hole expansion is mainly due to an enlargement process expe-

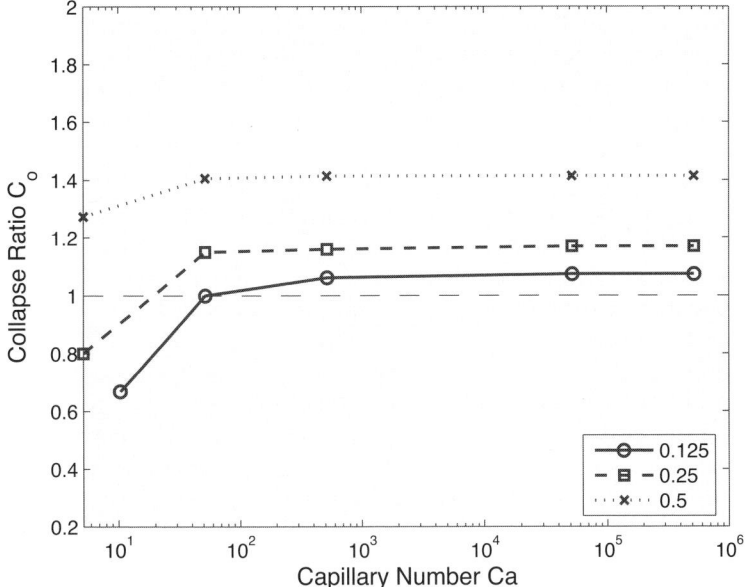

Fig. 6.2. Collapse ratio as a function of capillary number at different aspect ratios ($\varepsilon = 0.125$, 0.25, 0.5) for an initial radius ratio $\chi = 0.5$. After Xue et al. [2005b] (©[2005] IEEE).

rienced in the upper portion of the neck-down region. This insight has two consequences. Firstly that this phenomenon might well be exploited to counter hole collapse due to surface tension effects, and secondly that heat transfer taking place in the draw zone is a critically important fabrication parameter as this largely determines the length and shape of this deformation zone (see Section 6.3).

It is now known that the mechanism driving hole expansion relates to the force components arising in a non-slender deformation due to the application of draw tension [Xue et al. 2005b, 2006].

6.2.2 Hole-Shape Changes

As optical performance of any microstructured fibre depends strongly on the features of the microstructure, deformation during the draw process can potentially change the designed optical properties substantially. This makes understanding hole deformation critical. the complexity of this problem generally requires a computational approach. In cases where detailed numerical simulation is not an option, it is possible to estimate the maximum extent of hole size changes during drawing based on experimental data (e.g. see Table 5.3) and compensate for this at the preform fabrication stage. A more general understanding of the deformation process also makes it possible to develop some intuition about the likely type and extent of deformation.

Shape change is a complex phenomenon driven by interactions between neighbouring holes during the draw process. As such no relationship (as simple as Eq. 6.4 for C_o) seems possible even under isothermal drawing conditions. However, numerical simulations [Xue et al. 2005b] reveal some trends: surface tension is not a major factor and shape changes become more dramatic as the draw ratio is increased. For a fixed length, a higher draw ratio means a steeper neck-down region. Thus hole shape changes would seem to be closely related to the shape of the neck-down region which again points to the importance of heat transfer in the draw zone as a critically important fabrication parameter.

For given draw conditions, the extent of any shape change is influenced by both hole size and the spacing between holes. Changes become greater as the size disparity increases and the spacing is reduced. For different size holes, the shape changes may be characterised as a process whereby the hole with a larger curvature (i.e. the smaller hole) changes its shape in such a way that the curvatures of the two holes tend to match each other. Thus the smaller hole tends to change its original circular shape to that of an ovoid with its major axis parallel to the surface of the larger hole. This process is illustrated in Fig. 6.3 which shows the hole shape changes arising during a two-stage mPOF draw process for a five hole structure. A cross-section cut from the intermediate cane well away from the steep neck-down region (which promotes hole expansion) is shown in Fig. 6.3(a). Here the smaller holes have not dramatically changed their shape as a result of the initial draw process. For comparative purposes, Fig. 6.3(b) shows a cross-section cut from the final fibre drawn from the intermediate cane. Dramatic changes in hole shape are clearly evident with curvature matching deforming the small holes from near circular to decidedly ovoid.

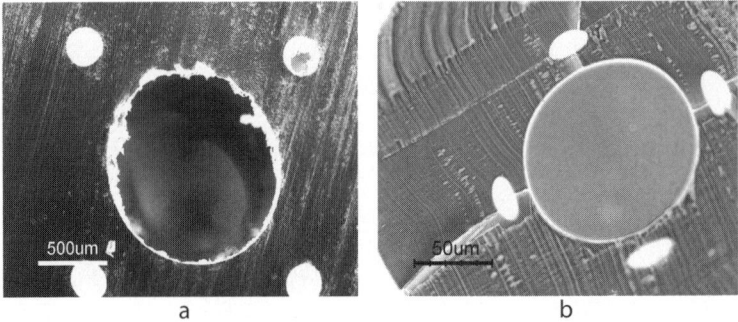

a   b

Fig. 6.3. Cross-section of (a) PMMA cane after drawing from primary preform, and (b) final fibre after the secondary draw. After Xue et al. [2005a] (©[2005] IEEE).

Interestingly, such deformation can be cascaded between neighbouring holes even if these all originally have the same size [Xue et al. 2005b]. Thus rings of small holes placed around a central larger hole will all deform. The innermost ring will deform most but the outer rings will also be deformed as their constituent holes change shape to match their curvatures to that of their deformed inner neighbours.

In many cases (e.g. GImPOF), the impact of modest hole shape deformation will not have a major impact on a fibre's optical performance. Indeed in some cases the interactions between different sized holes can be exploited, for example through the creation of elliptical holes that enhance birefringence effects [Issa et al. 2004]. However in one particularly important case, that of hollow or air-core fibre, hole deformation can present a major problem. Bandgap fibres intrinsically have at least two hole sizes - those defining the multi-ring cladding structure and the central core itself. When drawn, such a structure would be expected to significantly deform the inner ring of smaller holes (via curvature matching) with their deformation being cascaded onto the outer rings. Figure 6.4 shows the hole shape deformations that occur in both (a) a GImPOF (b) hollow-core mPOFs.

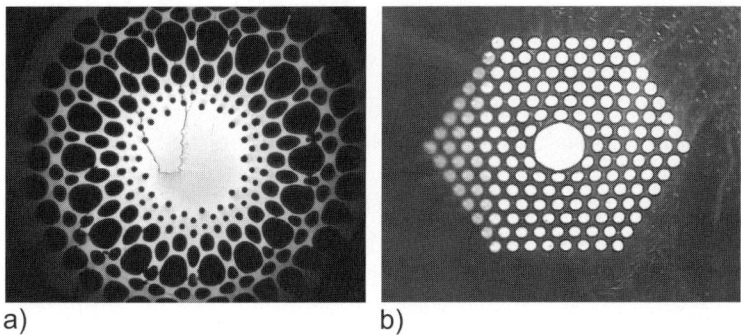

a)   b)

Fig. 6.4. Cross-sections through PMMA mPOFs (a) graded-index, and (b) hollow-core. After Xue et al. [2005b] (©[2005] IEEE).

Significant deformation can also occur when adjacent holes are the same size - here the key parameter is the hole spacing. As the spacing is reduced, so the interaction becomes more pronounced with the holes tending to deform so as to "merge" with each other. This behaviour can be seen in the suspended core fibre cross-sections in Fig. 6.9. Such deformation can work to the fibre fabricator's advantage, as Fig. 6.9(b) shows that the hole shape changes have resulted in longer, thinner material bridges between the holes. For this type of fibre, bridge thickness is the primary determinant of optical performance [Issa 2004].

6.2.3 Approximate Analysis Of Hole Behaviour

In the early days of mPOF fabrication, conventional wisdom was (i) that the drawing process would only ever lead to a reduction in hole size, and (ii) that surface tension was the dominant force in such hole deformation. Experimental tests and simulation have clearly established that hole expansion does occur in some conditions, and surface tension is only part of the deformation story.

The dominant forces in continuous fibre drawing (in the absence of hole pressurization) are draw tension and surface tension. Numerical simulations under isothermal drawing conditions lead to two observations. Firstly that hole expansion only takes place when the slender body assumption is invalid, and secondly that the extent of hole expansion is largest where the slope of the neck-down region is changing very rapidly. Both these insights can be readily explained by considering how an applied draw tension induces a viscous force F_{visc} around a hole [Xue et al. 2005b]. From a fibre fabrication perspective, the slender body assumption will never be completely met, and thus the question arises as to the relative importance of F_{visc} (which can lead to hole expansion) and the surface tension force F_{surf} acting around a hole (which always promotes hole contraction). The ratio of these two forces can be approximated by

$$\frac{F_{\text{visc}}}{F_{\text{surf}}} \approx 2 \left| \frac{dR_p}{dz} \right| Ca \qquad (6.5)$$

where Ca is the localized capillary number ($Ca \equiv \mu v_{\text{preform}}/\sigma$) with μ and σ being the local material viscosity and surface tension coefficient, respectively. This simple relationship reveals why quite different deformation results might be expected when drawing microstructured fibres made from different materials. Figure 6.5 shows the value of Ca for PMMA and silica over typical draw temperature ranges and at a realistic feed rate ($v_{\text{preform}} = 2.5$ mm/min). For PMMA, Ca varies over the range 10^2 to 10^6 while for silica the values are lower covering the range 1 to 10^3. It is this difference which lies at the heart of the very different types of hole deformation that occur with these two materials [Xue et al. 2006]. When Ca is large (as is generally the case with drawing PMMA to fibre) surface tension can actually be neglected at all but the highest temperatures encountered during the draw process. At lower Ca values (as is generally the case for silica drawing) surface tension plays a far greater role in hole collapse.

The second factor in Eq. 6.5 that influences the relative importance of the induced viscous and surface tension forces relates to the shape of the neck-down region, or more specifically, to the rate at which its radius changes with axial position. Although this analysis is strictly only applicable to isothermal drawing, this latter result has major implications for any fibre fabrication carried out under nonisothermal conditions, as what takes place in the neck-down region is strongly influenced by the heat transfer that occurs there and the impact that the rate of heat transfer has on the temperature profile (both in the axial and radial directions). Such issues are addressed in Section 6.3.

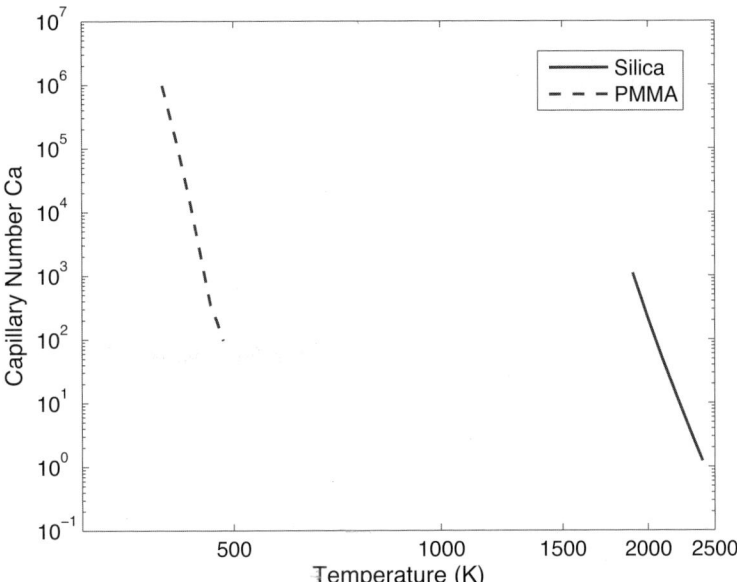

Fig. 6.5. Capillary number as a function of temperature for PMMA and silica. After Xue et al. [2006] (©[2006] IEEE).

Both the type of size change (expansion or collapse) and the rate of size change are affected by a hole's position in the preform. This determines both the local temperature (and thus the local physical properties) and whether or not the hole and its neighbours are in a region where preform radius is changing rapidly. As the sizes of adjacent holes change differently, so their size changes impact on the movement of material in their vicinity. Material may find itself pushed away from an expanding hole, pulled in towards a contracting hole, or more likely experience something more complex under the influence of a number of nearby holes. Shape changes are therefore produced by material movement induced by size changes.

6.3 Nonisothermal Analysis Of Hole Deformation

Useful as the insights from isothermal considerations are, practical fibre drawing is an inherently nonisothermal process and thus the following issues need to be addressed:

- How rapidly can the cane or preform be brought up to the necessary draw temperature?
- What is the magnitude of the axial and radial temperature (and thus viscosity) gradients during the draw process?
- What is the impact of the hole structure on internal heat transfer?

The previous chapter described two mPOF drawing systems. The first employed high intensity convective heating within its furnace (see Section 5.2.1), while the second used radiative heating (see Section 5.2.2). These two heating approaches are further examined here, in the light of their impact on hole deformation.

6.3.1 Convective Heat Transfer

Section 5.2.1 presents the design of a furnace (see Fig. 5.13) that has been successfully used in a drawing system for fabricating a wide range of mPOFs.

The need for an efficient preheater in such a system was established by examining the transient thermal response of two PMMA sections each 5 cm in diameter and 14 cm in length - one being a solid rod, the other having three rings of 2 mm diameter holes arranged in a hexagonal pattern [Lyytikäinen et al. 2004] similar to what might be used to create a single-mode fibre. Ten T-type thermocouples were embedded in each section to measure the temperature at various positions during the heating-up process which was provided by forced convection using hot (130-140 °C) air blown in via a ring of holes situated near the centre of a metal cylinder encasing the entire PMMA section. Figure 6.6 shows the measured temperature profiles across each section at different times from the commencement of heating (which was continued for a period of two hours). Despite good external heat transfer conditions, the low thermal conductivity of PMMA (together with the air within the microstructure) resulted in radial temperature gradients that were still significant (i.e. greater than 30 °C from centreline to wall) after 30 mins of heating.

Thus with a convective furnace it is important that sufficient time is allowed for the cane or preform to be brought up to temperature in a preheat section before fibre drawing is commenced, so as to minimise any radial temperature gradients that might distort the microstructure. The impact of a modest radial temperature profile (and thus a non-uniform viscosity profile) as a PMMA cane or preform leaves the preheating section and enters the draw zone has been shown by numerical simulation [Xue et al. 2005b] to have a minimal effect on the shape of the neck-down region.

Once the preform has left the preheat section, the challenge is not only to bring it up to its draw temperature but to do so with as small a radial temperature gradient as possible and with a neck-down shape that contributes as little as possible to deformation of the holes within the microstructure. Given the relatively low temperatures used here (the maximum being about 250 °C for the hot nitrogen stream), it might be felt that thermal radiation plays no part in this process at all. Certainly thermal radiation can be neglected both as a heat transfer mode to the external surface of the preform and within the bulk PMMA portion of the preform. However many mPOF designs contain extensive hole structures that effectively present an insulating air blanket around the central core region. Recent studies have shown that in

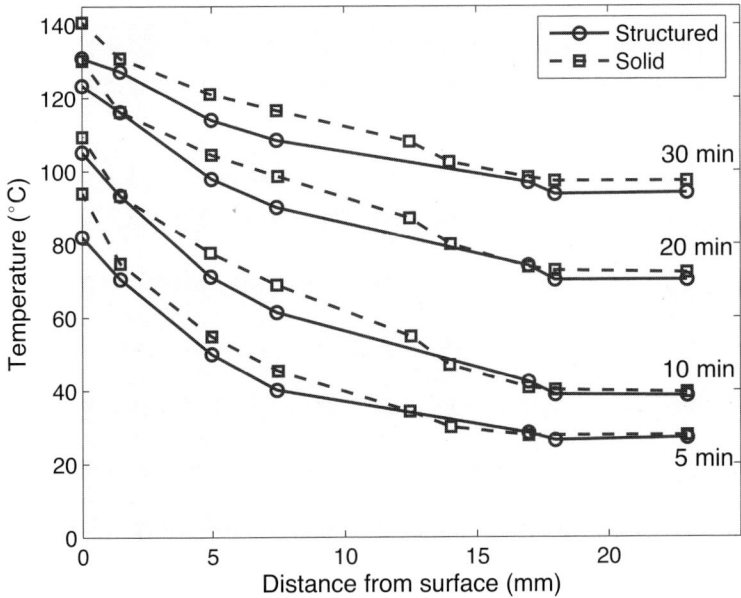

Fig. 6.6. Radial temperature profiles across solid and microstructured PMMA sections as a function of convective heating time. After Xue et al. [2007a] (©[2007] IEEE).

such cases, radiative heat transfer across the hole structure can be an important contributor to the total inwards heat flux [Lwin et al. 2006], while for silica MOF such radiative transfer plays a central role [Xue et al. 2007b].

Under steady-state conditions, the rate of convective heat transfer from the hot nitrogen to the external surface of the preform equals the internal conductive rate away from the surface,

$$\kappa \frac{\partial T}{\partial r} = -h[T - T_{N_2}(z)], \tag{6.6}$$

where T_{N_2} is the temperature of the nitrogen and h is the convective heat transfer coefficient. For continuous draw processes, h is a function of the axial position within the furnace due to variations in both diameter and axial velocity. For example, in this convective mPOF furnace, the external heat transfer can be adequately described by two heat transfer coefficients. In the upper portion of the neck-down region, where the cane diameter is essentially constant and the cane is moving slowly, an estimated value for h of 7.6 W/m²K was determined, based on local gas flow conditions within the furnace. In the lower portion of the neck-down region, where the fibre diameter is decreasing and the fibre velocity increasing, h may be estimated by

$$h = 128.3 \, v^{0.574} \tag{6.7}$$

where v is the local axial fibre velocity. Using this correlation gives an estimated value for h (under typical operating conditions) of 13.6 W/m²K which is nearly double that in the upper region. For comparative purposes and using the operational conditions listed in Table 6.1, the values for h when drawing silica fibres would be substantially higher (from 23 W/m²K in the upper part of the neck-down region to 191 W/m²K in the lower part), although these increases have only a minor effect on the drawing process as thermal radiation is the dominant heat transfer mode.

When drawing an mPOF, the magnitude of the radial temperature gradient is largely determined by the size of the local Biot number,

$$Bi = \frac{hR_{\text{preform}}}{\kappa} \tag{6.8}$$

which as noted previously is the ratio of internal and external resistances to heat transfer in the radial direction. At very low Biot numbers ($Bi \ll 0.1$), temperature, and hence the viscosity, is essentially uniform across any given cross-section. As Biot number increases ($Bi = 0.1$ to 1.0), internal conductive heat transfer becomes more important with a more pronounced radial difference occurring particularly in the upper portion of the neck-down region where the radius is still relatively large. However even in this heat transfer regime, the rapid thinning as the preform is drawn to fibre means that the local value of the Biot number (and the radial temperature gradient) rapidly decreases bringing a return to near isothermal conditions.

Figure 6.7 shows numerically calculated (using the PolyFlow package) temperature profiles for a solid PMMA fibre draw for five different values of Bi (0.01 to 1.0) corresponding to different h values (from 0.4 to 40 W/m²K). As expected, the radial temperature difference (i.e. centre-line to surface) for a given axial position increases with Bi. However by the end of the hot-zone (taken here as 50 mm long), this radial temperature difference between cases has essentially disappeared. This is consistent with the fact that the local Bi has dropped to significantly below 0.1 by the end of the draw zone with the result that radial differences in temperature, viscosity and axial velocity at this point have effectively ceased to exist.

The shape of the neck-down region under the same set of heat transfer conditions (i.e. for the same set of Bi values) is shown in Fig. 6.8, noting that the isothermal case has been included here for comparative purposes. Clearly the external heat transfer (defined here in terms of Bi) has a dramatic impact. However it is interesting to note that the onset of the neck-down region is delayed by increasing Bi between 0.01 to 0.1 but that this onset then starts earlier as Bi is increased between 0.1 to 1.0. The reasons for this behaviour are discussed in [Xue et al. 2005b].

Thus external heat transfer (set by the flowrate and temperature of the hot nitrogen) has a major role to play in microstructure deformation through the impact it has on the temperature and shape of the neck-down region. If the external rate of heat transfer is too high, then not only do internal

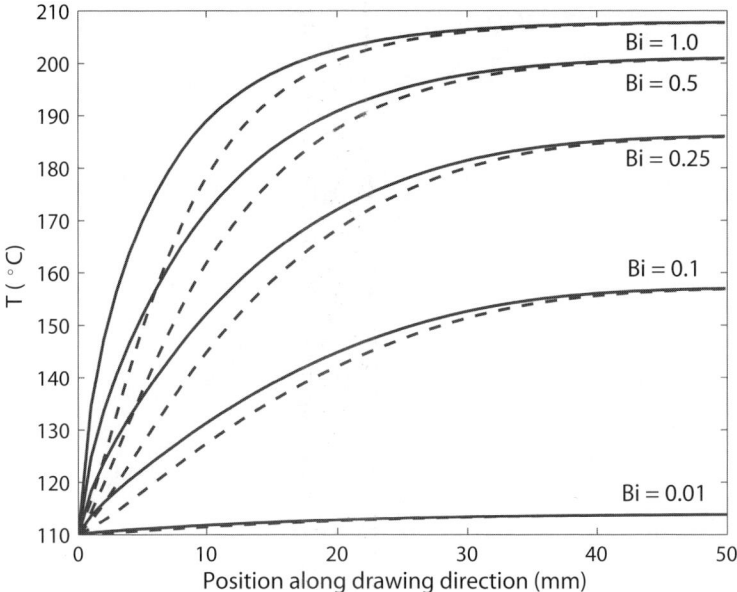

Fig. 6.7. Temperature profiles along the fibre surface (solid lines) and the fibre centreline (dotted lines) as a function of Biot number. After Xue et al. [2005b] (©[2005] IEEE).

temperatures rise rapidly to levels that are unsuitable for fibre drawing (see Fig. 6.7) but the start of the neck-down region (see Fig. 6.8) is characterised by a very rapid change in radius that may lead to excessive hole deformation (see Eq. 6.5).

6.3.2 Radiative Heat Transfer

One means of reducing the impact of poor internal heat transfer is to employ radiative (rather than convective) heating. This approach has been used very successfully (see Section 5.2.2). The radiative heating system employed not only brings the inner portion of the cane up to the drawing temperature more quickly but it also promotes a more uniform radial temperature.

As an example, Figure 6.9 shows cross-sections of two suspended core fibres [Lwin et al. 2005]. The fabrication process of both fibres was identical, except that one on the left was drawn using a convective furnace, while the one on the right was drawn using a radiative furnace. It is clear that the latter has produced a fibre where the material bridges between adjacent holes is noticeably thinner, resulting in a significant reduction in loss (see Table 6.3. The loss was reduced by more than a factor of 10 in going from a convective to a radiative draw furnace, while a further halving of the loss was achieved by refining the draw conditions in the later case. The major factor determining

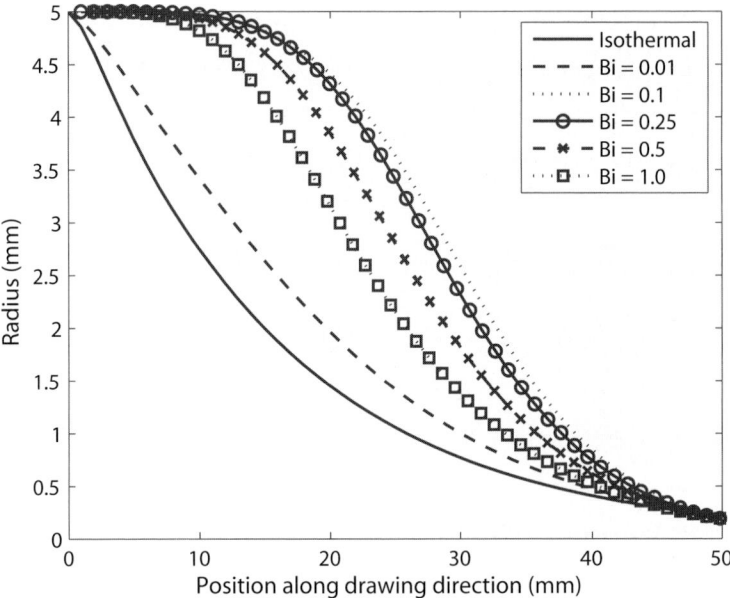

Fig. 6.8. Shape of the neck-down region for a solid fibre drawn for different Biot numbers. After Xue et al. [2005b] (©[2005] IEEE).

the loss performance is the thickness of the bridges between the holes, which control the confinement loss [Issa 2004].

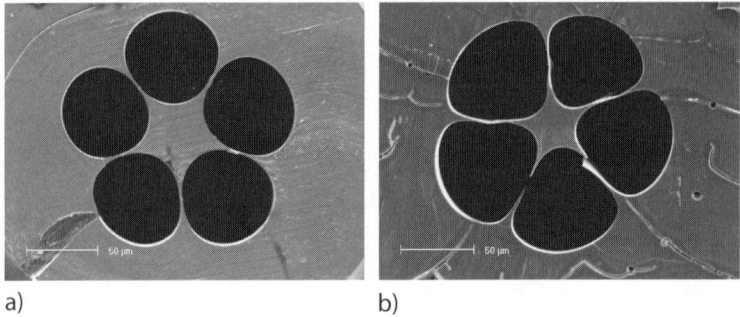

a)   b)

Fig. 6.9. Suspended core fibre drawn in (i) a convective furnace, and (ii) a radiative furnace.

6.3.3 Pressure Modified Hole Deformation

In Section 6.2.3, the point was made that hole deformation is determined by both the local forces acting around that hole and by local material movement

Table 6.3. Impact of furnace type on performance of suspended-core fibre.

Furnace Type	Bridge Thickness (μm)	Loss at 650 nm (dB/m)
Convective	1.79	3.85
Radiative ... (a)	0.62	0.36
Radiative ... (b)	0.55	0.19

0 mBar

8 mBar

10 mBar

13 mBar

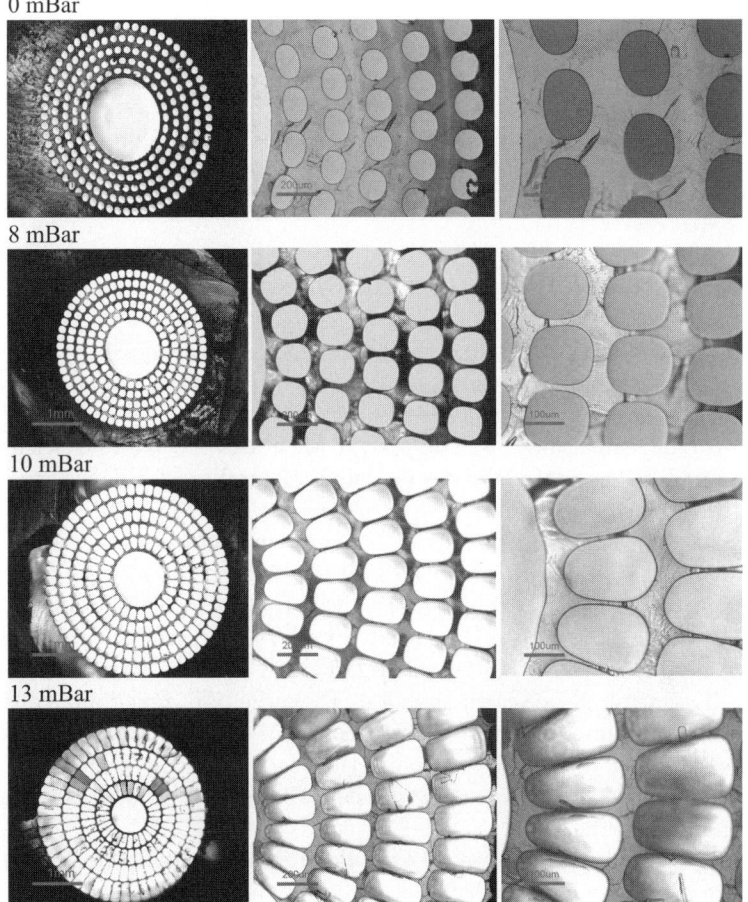

Fig. 6.10. Impact of internal pressurisation on hole structure for a range of cladding pressures.

brought about by the deformation of neighbouring holes. Thus far the focus has largely been on determining how best to create drawing conditions that minimise hole deformation. However the nature of the entire hole structure can also be manipulated by applying an internal pressure, thereby introducing an additional force that influences the behaviour of each hole. Overall the impact of applying such an internal pressure would be to counter any hole contraction and accelerate any hole expansion. Such an approach is not difficult to implement. However the required pressures are low and care needs to be taken to prevent holes "exploding" due to over-pressurization.

Such an option is particularly attractive when the holes pressurised are all the same size and thus asymmetries brought about via this force acting over different surface areas are minimised. Figure 6.10 clearly shows the dramatic effect that even a low level of internal pressurisation can have on a hole structure, and particularly on the thickness of the material bridges between adjacent holes. The thickness of these bridges plays a major role in the optical performance of this hollow-core Bragg fibre (see Section 3.1 and also Chapter 8).

References

Bansal, N P and Doremus, R H (1986). *Handbook of glass properties*. Academic, New York, USA.

Deflandre, G (2002). Modeling the manufacturing of complex optical fibres: the case of the holey fibres. In *Proceedings of the International Colloquium*, volume 2, pages 150–6, Valenciennes, France.

Fitt, A D F, Furusawa, K, Monro, T M, Please, C P, and Richardson, D J (2002). The mathematical modeling of capillary for holey fibre manufacture. *Journal of Engineering Mathematics*, 43(2):210–27.

Issa, N A (2004). High numerical aperture in multimode microstructured optical fibers. *Applied Optics*, 43(33):6191–7.

Issa, N A, van Eijkelenborg, M A, Fellew, M, Cox, F, Henry, G, and Large, M C J (2004). Fabrication and study of microstructured optical fibers with elliptical holes. *Optics Letters*, 29(12):1336–8.

Lwin, R, Barton, G, Keawfanapadol, T, Large, M, Poladian, L, Tanner, R, van Eijkelenborg, M A, and Xue, S (2005). Suspended core microstructured polymer optical fibre: Connecting to reality. In *Proceedings of the Australian Conference on Optical Fibre Technology*, volume 30, Star City, Sydney, Australia.

Lwin, R, Barton, G W, Large, M C J, Poladian, L, and Xue, S (2006). Heat transfer in preforms for microstructured polymer optical fibres. In *Proceedings of the International Plastic Optical Fibres conference*, Seoul, Korea.

Lyytikäinen, K, Zagari, J, Barton, G, and Canning, J (2004). Heat transfer within a microstructured polymer optical fibre preform. *Modelling and Simulation in Materials Science and Engineering*, 12(3):S255–65.

Myers, M R (1989). A model for unsteady analysis of preform drawing. *Journal of the American Institute of Chemical Engineers*, 35(4):592–602.

PolyFlow (2003). PolyFlow User's Manual, Ver. 3.10. Fluent Inc., Centerra Resource Park, Lebanon, New Hampshire.

Reeve, H M, Mescher, A M, and Emery, A F (2001). Experimental and numerical investigation of polymer preform heating. *Journal of Materials Processing and Manufacturing Science*, 9:285–301.

Reeve, H M, Mescher, A M, and Emery, A F (2003). Steady-state heat transfer and draw force for pof manufacture. In *Proceedings of the International Plastic Optical Fibres conference*, volume 12, pages 220–3, Seattle, USA.

Schultz, W W and Davis, S H (1982). One-dimensional liquid fibers. *Journal of Rheology*, 26(4):331–45.

Wu, S (1970). Surface and interfacial tensions of polymer melts. II: Poly(methylmethacrylate), poly(n-butylmethacrylate), and polystyrene. *Journal of Physical Chemistry*, 74(3):632–8.

Xue, S C, Large, M C J, Barton, G W, Tanner, R I, Lwin, R, and Poladian, L (2006). Role of material properties and drawing conditions in the fabrication of microstructured optical fibres. *Journal of Lightwave Technology*, 24(2):853–60.

Xue, S C, Lwin, R, Barton, G W, Poladian, L, and Large, M C J (2007a). Transient heating of PMMA preforms for microstructured optical fibres. *Journal Of Lightwave Technology*, 25(5).

Xue, S C, Poladian, L, Barton, G W, and Large, M C J (2007b). Radiative heat transfer in preforms for microstructured optical fibres. *International Journal of Heat and Mass Transfer*. In press.

Xue, S C, Tanner, R I, Barton, G W, Lwin, R, Large, M C J, and Poladian, L (2005a). Fabrication of microstructured optical fibres, Part I: Problem formulation and numerical modelling of transient draw process. *Journal of Lightwave Technology*, 23(7):2245–54.

Xue, S C, Tanner, R I, Barton, G W, Lwin, R, Large, M C J, and Poladian, L (2005b). Fabrication of microstructured optical fibres, Part II: Numerical modelling of steady-state draw process. *Journal of Lightwave Technology*, 23(7):2255–66.

7

The Handling and Characterisation
of Microstructured Polymer Optical Fibres

"From the moment I picked your book up until I laid it down I was convulsed with laughter. Some day I intend reading it."

Groucho Marx (1895-1977)

This chapter outlines three methods for cutting mPOFs as well as giving details of techniques that have been used to characterise mPOFs through accurate imaging of their microstructure as well as outlining measurements of their light guiding properties such as loss, numerical aperture, bandwidth and birefringence.

7.1 Cutting And Handling

An important issue with respect to both research and successful commercialisation is cleaving of a fibre to form an optical end-face suitable for coupling or connectorising. For conventional (all-solid) POF various cutting, polishing and connectorising methods and tools are available [Daum et al. 2002, Rennsteig 2005]. However, these are not very effective when used on mPOFs due to their non-standard diameters and the necessity to keep the air-hole structure open at the fibre end-face. Several mPOF-specific alternatives are discussed in this section, namely melt, laser and hot-knife cutting. The latter is currently the preferred method and will be treated in greater detail.

7.1.1 Melt Cutting

Polymer fibres can be cut by bringing the fibre close to a localised heat source (such as the tip of a soldering iron) whilst providing tension to the fibre. This will result in local softening of the fibre, followed by necking down and breakage. The resulting fibre end generally has a funnel-shaped section of a few millimetres in length at the tip, with the fibre diameter at the end-face being significantly larger than the original fibre diameter (thought to be the result of a partial release of the "frozen-in stress" from the fibre draw). The surface of the end-face is concave (inward facing) to a depth of a few microns (see Fig. 7.1). Due to the likelihood of asymmetric heating, the resulting fibre also often has a short bend over 10 to 30° just before the end which causes light to emerge from the fibre with a slight transverse offset, distorting the symmetry of near- and far-field patterns. Despite these problems, the melt cutting method has been used successfully to prepare fibres. The funnelled "expanded core" section can be used to enhance the efficiency of launching into small cores and the surface of the end-face is particularly smooth though the structure is slightly altered by the melt-cutting process.

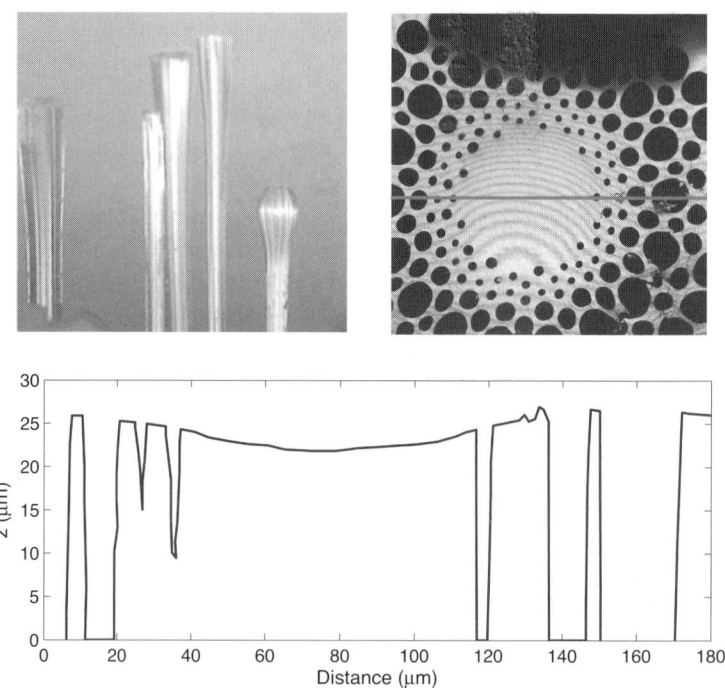

Fig. 7.1. Examples of the cut ends of fibres obtained using melt cutting (top left), the resulting transverse cross section of a melt-cut GImPOF (top right) and tomography (bottom) of the cut surface showing the concave enface. *Data courtesy of the Polymer Opitcal Fiber Application Centre, POF-AC Nürnberg, Germany.*

7.1.2 Laser Cutting

Laser ablation can generate reasonably good cleaved ends of mPOFs in seconds. This has been demonstrated with an ArF Exciplex UV laser operating at 193 nm with a pulse width of 20 ns. It was found that laser cleaving at a 4 Hz repetition rate and with an intensity of 1.6 J/cm^2 produced satisfactory cleaves [Canning et al. 2002]. The pulsed nature of the cutting was observed to result in a slightly saw-tooth like surface. Cost, safety and portability issues for such a laser system, however, are obvious problems for any practical application making it unlikely that this method could be of any general use in the field.

7.1.3 Hot-Knife Cutting

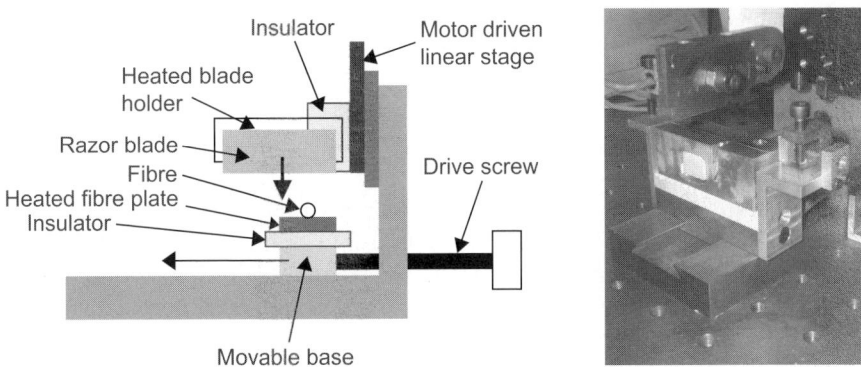

Fig. 7.2. Automated mPOF cutter device employing a heated base and motor driven razor blade. After Law et al. [2006b].

Figure 7.2 shows a rig specifically designed to investigate the mPOF cutting process. The fibre is clamped in a V-groove on a heated platen with a cross-groove to prevent collision damage to the heated blade. In addition to independent control of blade and platen temperatures, the blade is mounted on a stepper-motor driven linear stage to enable both speed and cut depth to be controlled. Scoring on the end-faces of mPOF suggests that grooves on the blade (and blade damage from previous cuts) are responsible for some of the damage to the end-face of a cut fibre. Damage can also be caused by PMMA debris being stuck to the blade and being drawn across the end-face. Therefore a method of translating the blade relative to the fibre by a screw-driven slide is implemented to ensure that a "virgin" section of blade is used each time. A 1 mm cross-groove is provided to allow the blade to pass completely through the fibre whilst minimising the length of unsupported fibre. Fibres are left to thermally equilibrate on the platen for at least 60 s before cutting.

It is possible to consistently cleave mPOFs to produce a good quality optical end-face using this rig. As an example, we cut a graded-index mPOF (or GImPOF) with an outer diameter of 400 μm that had been drawn at a temperature of 220 °C and under 40 g of tension. Before cutting, the outer

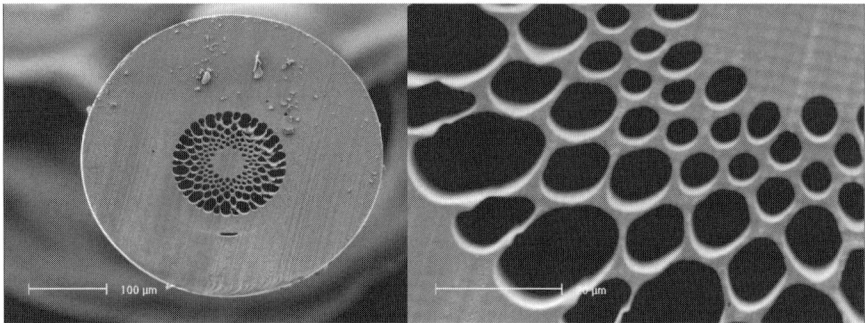

Fig. 7.3. SEM image of a 'good' cut of a graded-index mPOF achieved with a blade temperature of 80 °C, a platen temperature of 70 °C and a blade speed of 0.07 mm/sec.

surface is cleaned with isopropanol. Figure 7.3 shows a scanning electron microscope (SEM) image of the fibre cut. The cutting conditions are a fibre (platen) temperature of 70 °C, a blade temperature of 80 °C and a blade speed of 0.07 mm/sec. The cut was made from the upper right to the lower left as shown, with a surface characteristic of a ductile cut with lines marking the progress of the blade across the surface. Faint lines perpendicular to the cutting direction are visible in the close-up image in both the core and cladding. These lines are 2 to 2.5 µm apart, a separation that suggests their origin is the 2.03 µm movement of the stepper motor that drives the blade. The lowest portion of the cut surface corresponds to more of a ductile fracture with "river" lines showing the path of the fracture edge. There is some distortion of the (micron-thin) inter-hole walls at the top of the structure. Note that the extra crescent shaped "air hole" below the structure is actually a result of imperfect sleeving of the cane during the overall fabrication process.

High-quality mPOF cleaves can also be obtained with a well-tuned device employing a razor blade [Law et al. 2006a,b]. It was found that the important parameters for a good cut are: (i) the temperature of the cutting blade, (ii) the temperature of the platen holding the fibre, (iii) the time allowed for thermal equilibration between fibre and platen, (iv) the blade speed and (v) the blade condition. Optically acceptable mPOF end-faces can be achieved over a limited range of conditions. It is important to note, however, that an elaborate cutting setup as shown in 7.2 is not necessarily required to obtain good cuts. MPOFs with a thick cladding around a microstructured area of relatively low air fraction can be cut adequately by hand with a razor blade at room temperature, especially when the fibre has been drawn at low tension (i.e. with a low degree of frozen-in stress).

Role of material properties in cutting behaviour

Insights into the cutting behaviour of mPOF can be gleaned from an understanding of the material properties. The amorphous PMMA has a viscosity
molecular mass (M_v) of 7.2×10^4. The syndiotactic content (i.e. the fraction of
the polymer side groups that come on alternating sides of the backbone chain)
is about 65% [Mark 2004]. PMMA is usually regarded as a "glassy" polymer,
it is an optically clear thermoplastic that is hard at room temperature but
which can be worked visco-elastically above its glass transition temperature.
PMMA can, however, can be anomalous in its mechanical behaviour. In particular it can be significantly tougher/stronger than other glassy polymers.
It has been shown that the brittleness of bulk PMMA is strongly dependent
on its M_v value with brittle behaviour occurring for $M_v < 10^4$ and ductile
behaviour for $M_v > 10^5$ [Kusy 1977]. In addition, a "brittle to ductile" transition can occur when samples are heated to around 60 °C, a level well below
the glass transition temperature [Matsushige et al. 1976]. This behaviour has
been interpreted as evidence of a phase transition. In the brittle region, the
fracture mechanics are complex and molecular weight dependent [Kusy and
Turner 1977] involving the formation of "crazing cavities" ahead of the fracture tip [Passaglia 1987]. The fracture surface energy also exhibits complex
behaviour with respect to crack speed [McCrum et al. 1997] which can affect
the surface quality arising from any induced fracture. In the ductile region,
interfacial shearing plays a major role in determining the properties and deformations of the exposed surface.

The process of cleaving a material involves the production of a pair of
fracture surfaces either by cutting with a sharp edge or by propagation of a
crack. Which process occurs depends upon the properties of the material being
cleaved and the process parameters. For a hard, brittle material cleaving is
usually produced by crack propagation. In ductile or elastic materials, cracks
do not propagate as readily and the rupture is propagated by a sharp tip or
blade. If the material is very ductile, small polymer lumps can be dragged out
of the surface by the blade causing a smearing effect over the cut surface [Law
et al. 2006a]. The quality of the fracture surface is also material dependent.
The properties of glassy polymers, in particular the role of chain alignment
and crazing in crack formation and the subsequent fracture mechanics [Kausch
1978, Passaglia 1987], have meant that polymer optical fibres must usually be
either mechanically or thermally polished after cleaving at room temperature.
For mPOFs, even in the rare event of the core structure surviving this (cold)
cleaving, the subsequent polishing processes are likely to cause more damage
than improvement, and may introduce particulates into the holes.

The bulk characteristics of a material can be significantly modified by the
fibre fabrication process. It has long been known that viscoelastic drawing
of polymers creates anisotropy in the mechanical properties. In particular,
the longitudinal fracture surface energy (a measure of the material's resistance to cracking) is reduced while the transverse fracture surface energy is

increased [Bucknall 1977]. Quantitative data are limited to draw ratios very much smaller than those encountered in mPOF fabrication. Anisotropic behaviour and the microstructure together form a classic crack-stopping structure [Gordon 1991] where the stress concentrations at the crack tip are only able to form any significant crazing perpendicular to the crack direction. The result is that cracks arising from attempted cleavage turn through a right angle, causing splintering. The degree of anisotropy decreases with increasing draw temperature and less splintering would be expected in such cases. Clearly any generic method for cleaving mPOF must allow for this variation in material properties. Fortunately the key mechanical properties are temperature dependent. An increase in temperature typically reduces both the brittleness and the anisotropy of the fracture toughness [Kusy 1977, Matsushige et al. 1976].

7.2 Fibre Characterisation

This section outlines various techniques that have been used for the characterisation of mPOF. These are generally a combination of techniques used for the characterisation of silica photonic crystal fibres [Knight et al. 2001] and those used for conventional POF [Daum et al. 2002].

7.2.1 Structural Characterisation

Microstructured optical fibres owe their optical properties to the exact dimensions of their microstructure. This structure can be inspected in the first instance with an optical microscope to get confirmation of the overall structure and to provide an estimate of the structural dimensions. Light microscopes, however, have limited resolution and features comparable or smaller than the wavelength of light cannot be determined accurately. In addition, care must be taken with the method of illumination: when acquiring images of the fibre structure with illumination from below (i.e. through the fibre), the hole sizes will appear significantly larger than with illumination from above (i.e. in reflection off the fibre surface). In addition, fibre cleaves are not always perpendicular to the fibre and there can be a slight tilt in the sample mounting so that it can be difficult to get the entire fibre surface into sharp focus.

To obtain more accurate measurements of the fibre structure, electron microscopy is required. The procedure is as follows. One centimetre pieces of mPOF are cut and inspected with the optical microscope. These are then mounted with double-sided tape on the side of an SEM sample stub that has been milled in half to provide a vertical surface. For stability, the fibres are allowed to only stick out a few millimetres above the horizontal surface of the stub. Conductive carbon paint is applied to cover the samples, except for the tips which are coated with a 10 nm thick coating of either platinum or gold. The stubs are then placed in the SEM machine and relatively low

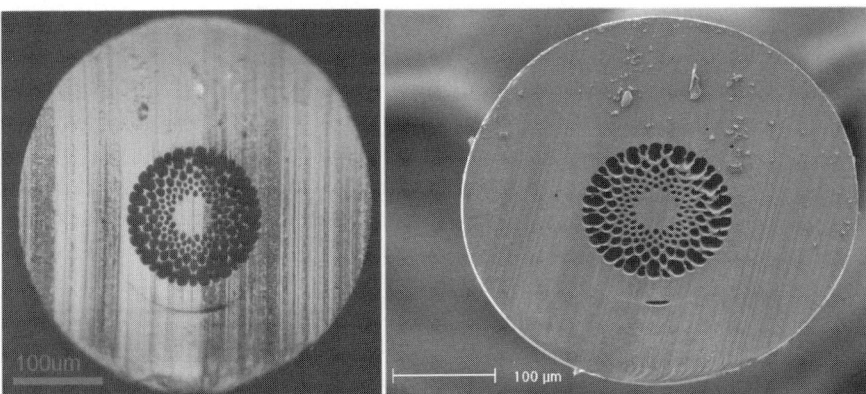

Fig. 7.4. Comparison of an optical microscope image (with top illumination) and a Scanning Electron Microscope image (right) of the same GImPOF sample.

voltages (≈ 5 kV) are used for imaging to avoid charging. The differences in image quality obtained with optical and electron microscopy are illustrated by the examples in Fig. 7.4, which were taken of the same fibre.

7.2.2 Single- And Multimode Guiding

The number of guided modes that are present in an optical fibre is an important aspect of both their design and their characterisation. The number of modes that are guided (or effectively guided) in a microstructured fibre generally depends on the wavelength of the light, the core size and on the ratio of hole diameter d to hole spacing Λ. For the most common hexagonal microstructure, a d/Λ smaller than 0.406 provides "endlessly" single-mode guidance [Kuhlmey et al. 2002]. Microstructured fibres however have no strictly bound modes in the same sense as a conventional step-index fibre with a core of elevated refractive index (see Section 3.2).

In practise, to investigate whether a fibre supports only a single mode, the following two experiments can be conducted [van Eijkelenborg et al. 2001]. Firstly, laser light (e.g. a HeNe laser of 632.8 nm) is launched into the MOF by butt-coupling to a multi-mode silica fibre or by launching with a high numerical aperture lens. Cladding modes are stripped off by putting a fluid with a higher refractive index than the fibre material on the fibre surface ($n > 1.5$ for PMMA). Alternatively, an extremely convenient way to strip cladding modes off mPOFs is to run a section of the fibre through very fine sandpaper. This gently scratches the fibre surface giving it a matt finish and the resulting scattering means that the cladding modes are quickly lost. A benefit of this latter method is that it does not require the use of index matching oils or liquids. After the cladding modes are stripped off, the fibre output is observed.

If the shapes of both the near- and far-field patterns are independent of the launch conditions and independent of any bends or twists in the fibre, then the fibre supports only a single mode. If a small number of modes were being guided, the near-field pattern would show modal interference lines and in case of a highly multi-mode fibre the output patterns would be speckled. In these latter two cases, the output patterns would also be extremely sensitive to disturbances of the fibre (i.e. bends, twists and launch conditions). The second approach involves a spatial interference measurement that can be performed between the collimated output of an MOF and the output of a standard single-mode glass fibre. If the MOF is single-moded, a clear interference pattern can be obtained with a very high fringe visiblility (achieved by balancing the power in the two arms). A translation of the fringes will be observed if there are thermal and mechanical disturbances.

7.2.3 Near- And Far-Fields

Near- and far-field patterns of the guided modes in MOFs (see Section 2.2.2) have been studied since their early days [Knight et al. 1996].The near-field pattern is recorded by imaging the fibre end-face onto a CCD camera and reducing light levels to avoid saturation. The far-field pattern is produced by removing the lens, and examining the output a few centimetres from the end of the fibre.

In typical hexagonal array MOF, light is generally strongly confined to the core region and the near-field pattern is dominated by minima occurring at the six nearest air holes. The far-field also has hexagonal symmetry with six symmetrically placed spots occurring around a central peak [Mortensen and Folkenberg 2002]. Near- and/or far-field distributions for mPOFs have been reported by many groups as a first means of characterising the fibre properties [van Eijkelenborg et al. 2001, Choi et al. 2001, Park et al. 2002, Choi et al. 2003, Huang et al. 2003, Kuzyk 2003, Asnaghi et al. 2002, Kondo et al. 2004, Huang et al. 2004b,a, Shin et al. 2004, Goto et al. 2004]. For mPOFs with hexagonal hole arrangements they were found to be virtually identical to those reported for silica MOF. Near-Gaussian transmission profiles were observed for terahertz radiation (radiation with a wavelength of the order of 1mm) guided in a Teflon MOF using a knife-edge measurement [Goto et al. 2004].

The modes of MOFs with non-hexagonal hole distributions have also been studied. Included here is the example of a rectangular "suspended" core mPOF with an aspect ratio between the major and minor core diameters of 9 : 1, which shows markedly different near- and far-fields. Scanning electron microscope images of this fibre are shown in Fig. 7.5 where the major axis of the core is 144 μm and the minor axis is 16 μm . The bridge thicknesses δ in the structure is on average 1.1 μm.

The near-field and far-field patterns of the fibre in Fig. 7.5 are shown in Fig. 7.6. The near-field showed strong speckle that was very sensitive to fi-

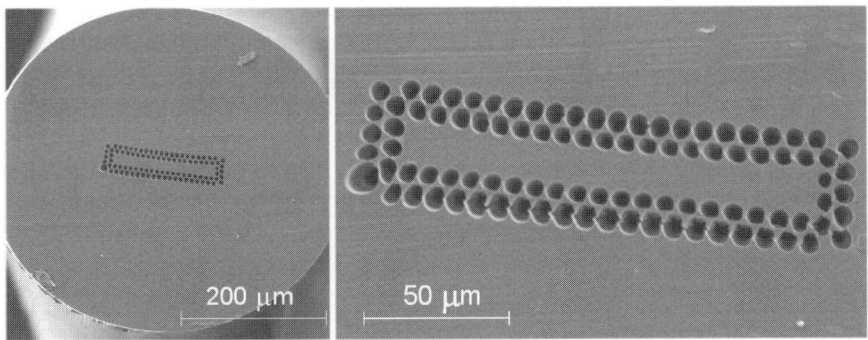

Fig. 7.5. SEM images of a rectangular core mPOF with core dimensions of 144 μm × 16 μm. After van Eijkelenborg et al. [2006].

bre bends and launching conditions, reflecting the multimode nature of the guidance. Attempts were made to only excite certain mode-groups by launching off-centre but for all fibre lengths down to a 20 cm length the near-field continued to show a completely filled (speckled) core. This indicates that the inter-mode coupling is very efficient; the equilibrium length for power distribution across the transverse modes is very short so that variations in the launch conditions have no visible effect on the output. The far-field patterns were circular [van Eijkelenborg et al. 2006].

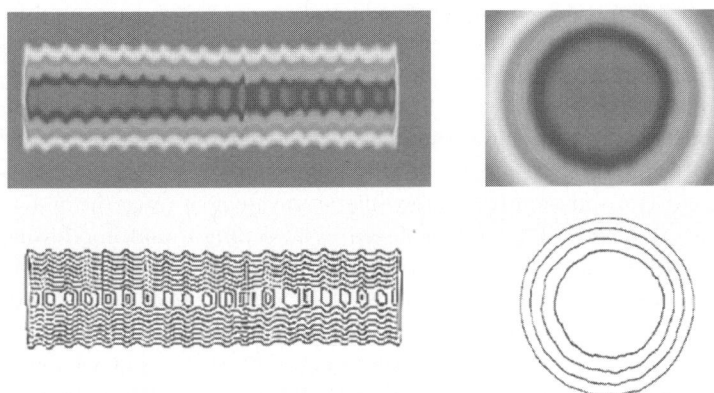

Fig. 7.6. Measured near- and far-field intensity distributions for a rectangular core mPOF. Note that the near-field images have a 2× scaled up minor axis for increased clarity. After van Eijkelenborg et al. [2006].

7.2.4 Numerical Aperture

Multi-mode optical fibres with wide light acceptance angles and high trapping efficiency are essential for a variety of applications. In recent years, such fibres have become increasingly important in such fields as the detection of charged particles and ionising radiation [Achenbach and Cobb 2003], multi-object spectroscopy [Mediavilla et al. 1998], beam-shaping, light delivery and most notably for cladding-pumped fibre lasers [Sahu et al. 2001, Bouwmans et al. 2003, Tunnermann et al. 2005]. Their application in short-haul communication has a long history due to their ease of connectivity and high launching efficiency. More recently their usefulness in the interconnection of electronic systems over very short distances has been recognised [Mortensen et al. 2003].

The highest numerical aperture fibres reported to date are MOFs of the suspended-core type, where a large core is virtually suspended in air by a system of very thin bridges that connect to an outer cladding. For such fibres, the effective core/cladding index contrast approaches that of the material/air contrast, and NA values of the order of 0.9 have been reported [Wadsworth et al. 2004b]. The NA is determined by the coupling between the core modes and the modes in the supporting bridges, which makes the bridge thickness the dominant parameter in determining the NA [Issa 2004].

In order to quantify the capture of light from large area, broad-beam sources, or the trapping of light emissions from within the fibre core, the nominal numerical aperture is generally defined for optical fibres as $NA = \sin \theta_{max}$ where θ_{max} is the maximum angle at which a meridional ray entering the fibre will be guided (note that a meridional ray passes through the optical axis of the fibre, in contrast to a skew ray, which does not). It is measured from the far-field angular intensity distribution by determining the half-angle at which the intensity has decreased to a (standard) 5% of its maximum value [Franzen et al. 1989].

For mPOFs, NA measurements have been taken by launching white light with a $60\times$ microscope objective ($NA = 0.8$) into the fibre and recording the far-field transmission intensities in the wavelength range from 450 nm to 850 nm at all angles. This was achieved by sweeping a multimode silica fibre connected to an optical spectrum analyser in an arc about the end-face of the microstructured fibre at a distance of 10 cm (far-field) and recording the transmission spectrum at increments of $1°$. The solid angle of detection is thus constant for all angles and wavelengths. The three-dimensional data set of transmission T, wavelength λ and angle θ (shown for example in Fig. 7.7) is then reduced to the two-dimensional data set of NA versus λ by using the 5% criterion for each angle.

It has been shown that for a circularly symmetric highly multimode fibre (such as the mPOF shown in Fig. 7.8) the NA strongly depends on bridge thickness and that an exceptionally high NA can only be achieved for bridge thicknesses much smaller than the wavelength [Issa 2004]. In Fig. 7.9 the wavelength dependence of the NA is plotted for three fibres with different structural

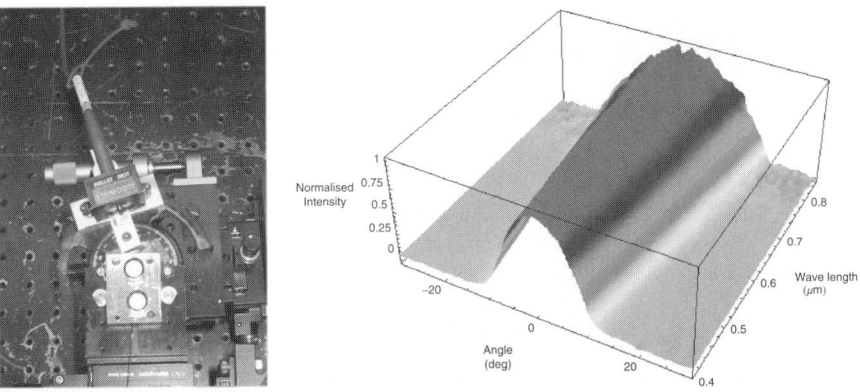

Fig. 7.7. Experimental setup (left) for measuring numerical apertures by collecting transmission spectra with a fibre connected to a spectrum analyser for all angles. The 3D raw data of transmission versus angle and wavelength (right) can be analysed to determine the *NA* versus wavelength.

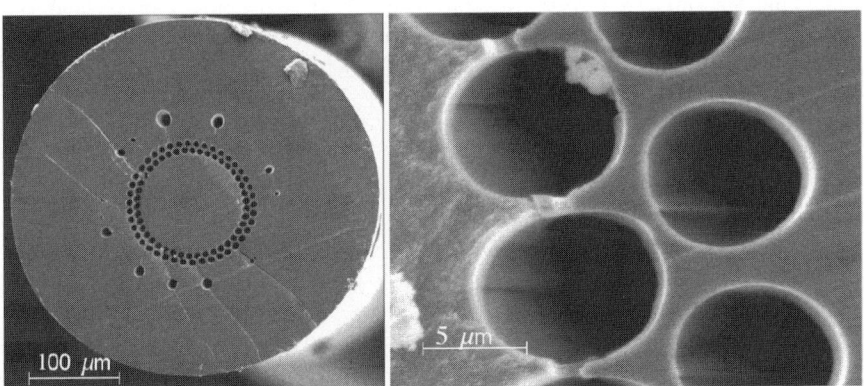

Fig. 7.8. Scanning electron microscope images of a circular-core mPOF with a bridge thickness of 0.99 µm.

dimensions. This was done by plotting the inverse scaled wavelength (δ/λ) for each fibre. A comparison of the measured *NA* with the theoretical predictions from Issa et al. [2005] shows good agreement over a broad wavelength range, with the highest *NA* measured being just over 0.5 at 850 nm for the mPOF with the smallest bridge thickness (0.66 µm).

7.2.5 Transmission Loss

A critical parameter for any optical fibre is its transmission loss. Fibre transmission loss is traditionally measured using the cut-back technique; by comparing the power transmitted through a length L of fibre with that transmitted

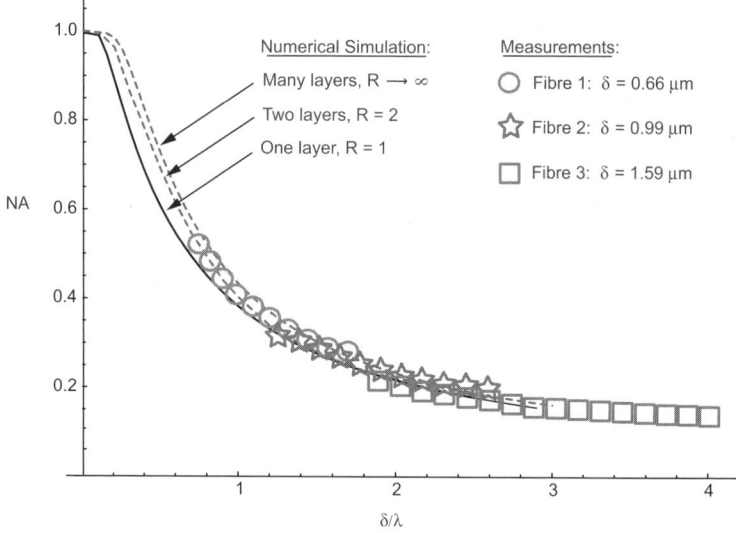

Fig. 7.9. Measured NA versus inverse scaled wavelength (δ/λ) for circular core mPOFs. The drawn curve shows the numerical calculations for a fibre with one, two, or many layers of holes. After Issa and Padden [2004].

through a cut-back piece of length $L - x$, the loss in the section of length x can be determined [Daum et al. 2002].

There are a few issues to keep in mind when performing such loss measurements. Firstly, the numerical aperture of the launch can be important, especially for short fibres, as over-fill or under-fill conditions can lead to higher and lower loss measurements, respectively. These effects are easily avoided when the inter-mode coupling is very strong; any launch NA is quickly redistributed to the equilibrium mode distribution, making the fibre output independent of the launch conditions after a short length of fibre. Secondly, the cladding modes need to be stripped off because the detector is often arranged to collect all light from the fibre end, including any light in the cladding. This should be done both directly after launching into the fibre as well as at the end of the fibre to account properly for confinement loss [White et al. 2001]. Thirdly, care must be taken to ensure that the power detection method captures all light emerging from the fibre core. This often requires some mode matching (NA and spot size). Finally, any effects of fibre end-face cutting quality on power transmission can be averaged out by recording the transmission power through several cuts at each measurement length.

Generally mPOF loss reported to date has been measured by launching a light source into at least 5 m of mPOF with a microscope objective, stripping off the cladding modes and coupling the fibre output to the input port of an optical spectrum analyser (OSA). The light sources used include lasers (for single wavelength measurements) or broadband sources such as a bright halogen 'white' light source or a super-continuum source. The latter is by far the most convenient for all broadband measurements as it is an in-fibre source with both a very high bandwidth and output power [Wadsworth et al. 2004a]. Detection is either through imaging the fibre end-face onto the detector diode of an optical spectrum analyser (OSA) using a low-magnification microscope objective, or by butt-coupling to a high-NA multimode acceptance fibre of a handheld OSA (e.g. an Ocean Optics USB4000). The fibre is then cut, either forward or backwards, in steps of 0.5 or 1 m, and the recorded transmission spectra are normalised to the shortest length (which is never shorter than 1 m to ensure sufficient mode mixing to avoid effects of over- or under-fill launch conditions).

Alternative methods of measuring fibre transmission loss such as Optical Time Domain Reflectometry (OTDR) are available, though they have not yet been successfully demonstrated for mPOF.

Causes of mPOF transmission loss

There are a number of causes of mPOF transmission loss, and much work has been devoted to understanding, isolating and eliminating them. These include material absorption, scattering, confinement, microbending and structural variation losses. These will each now be considered.

A number of polymer materials are available for fibre drawing including e.g. polycarbonate, polystyrene and Teflon, each with varying degrees of transparency and processability (see Chapter 5). The polymer most commonly used for mPOF fabrication is polymethylmethacrylate (PMMA) due to its high transparency (compared to the other commonly available polymers), relatively low cost, ease of processing and its inherent resistance to water and various oils [Daum et al. 2002]. The optical transmission loss of PMMA, however, is limited by the absorption due to the higher harmonics of the C-H bond vibrations (see Section 1.1 and Fig. 8.1 for PMMA loss data).

Material scattering can be split into (i) Rayleigh scattering which is an inherent characteristic of the material, and (ii) scattering due to dopants added to the basis polymer. Spatial fluctuations in dopant concentration cause scattering which can be a significant loss contributor for POFs. Studies conducted on CYTOP® [Onishi 2001] showed that doped fibres have scattering losses of 41 dB/km (out of a total reported loss of 47 dB/km) at 650 nm. Dopants are not, however, an issue for mPOFs as the microstructure provides the optical guiding as opposed to the dopant profile in conventional POF.

Material impurities can be unintentionally introduced into the material either during the polymerisation stage (e.g. in the case of preform casting) or during the extrusion/forming process. Small foreign particles can often be seen embedded in commercially extruded rods, visible as tiny black dots. Because of their relatively large size, they can cause significant wavelength-independent scattering loss. Material impurities can also be introduced by the methods used in the mPOF fabrication process. For instance, the choice of coolant used for the CNC preform drilling process can make a significant difference; it has been found that switching from an oil-based emulsion to a semi-synthetic cutting fluid provided a 28% loss reduction [van Eijkelenborg et al. 2004].

The hole structure in a drilled primary preform needs to be thoroughly cleaned out by extensive water flushing (without using organic solvents) to remove residual cutting fluid and any PMMA micro-swarf (an analysis of the cutting fluid after drilling showed PMMA microparticles of the order 0.1 to 2 μm diameter were present in the cutting fluid) [Barton et al. 2003]. Several preform flushing processes have been tried. Flushing out the preform hole structure with running tap water for ten minutes directly after the drilling process and keeping the preform in a dust free environment brought about a 40% reduction in the mPOF loss [van Eijkelenborg et al. 2004]. Similar results were obtained by flushing with water in a closed circuit for 3 hours. Attempts to enhance the cleaning process by sonicating the preform for 10 min in a hot water bath (which might be expected to reduce transmission loss by loosening drilling contaminants within the hole structure) did not bring about any improvements; this in fact doubled the loss which is thought to be related to the formation of micro-cracks in the polymer.

The impact of various options in the fibre production process on the loss for a GImPOF is summarised in Fig. 7.10. The GImPOFs labelled [A-F] were fabricated under different conditions as listed in Table 7.1. The PMMA bulk material loss curve [G] is included for reference, showing strong absorption peaks near 725 and 900 nm.

A sleeving process to enable drawing fibre with an external diameter that is up to 10× larger for given microstructure dimensions was developed (see Chapter 5). Although this was initially done to increase the ability to cut the fibres more easily, it was found to also significantly reduce the transmission loss, as shown in Fig. 7.11, which was attributed to a significant reduction in microbending effects [Nielsen et al. 2003, van Eijkelenborg et al. 2004].

Table 7.1. Processing options for various graded-index microstructured polymer optical fibres as presented in Fig. 7.10.

Fibre label	Cutting Fluid	Water Flushed	Sonicated	Sleeved
A	Regular	No	Yes	No
B	Regular	No	No	No
C	Regular	Yes	No	No
D	Regular	Yes	No	No
E	Regular	Yes	No	Yes
F	Synthetic	Yes	No	Yes

The lowest loss was obtained with fibre [F] which was fabricated with the synthetic cutting fluid, with preform water flushing, by avoiding sonication and by sleeving, producing a loss of 0.80 dB/m at 760 nm and 0.87 dB/m

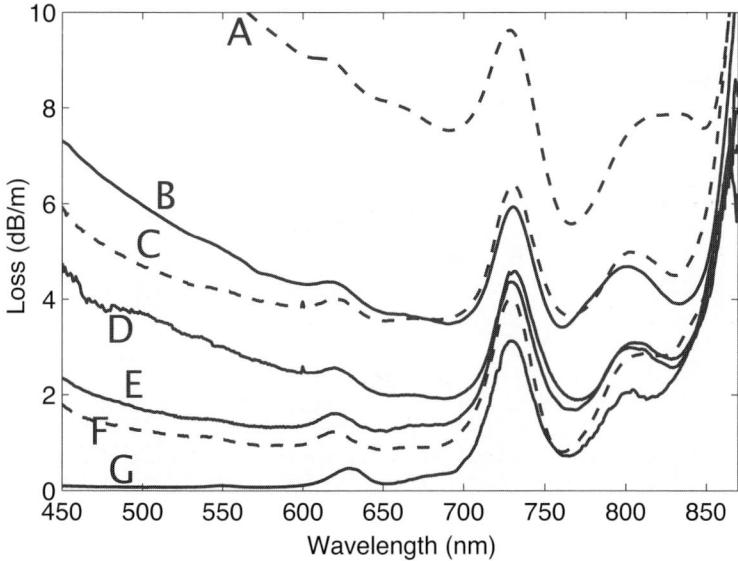

Fig. 7.10. Measured transmission loss for graded-index polymer optical fibres labelled [A-F] as fabricated under the conditions listed in Table 7.1 to determine the relative contributions of various fabrication procedures. Bulk material loss for PMMA [Daum et al. 2002] is indicated by curve G. After van Eijkelenborg et al. [2004].

at 650 nm (with the material loss at those wavelengths being 0.72 dB/m and 0.15 dB/m, respectively).

Figure 7.11 shows the measured loss for various mPOF designs as a function of fibre diameter (all drawn under similar conditions). Allowing for some scatter due to design differences, the results show the loss decreasing with diameter, no matter what the design [Lwin et al. 2005b, Large et al. 2006]. This trend indicates that microbending remains a significant loss contributor even for 500 μm diameter fibre. Further improvement should be possible with larger diameter fibres. Considering that conventional POFs are generally drawn to a diameter of nearly 1 mm, extrapolation of the mPOF results to this diameter indicates that mPOF loss could reach about 160 dB/km. Jacketing the fibres could bring about further loss reductions.

Non-adiabatic longitudinal variations in the transverse fibre structure can also cause loss. This is a major issue for designs where thin bridges separate relatively large holes, as these have optical properties that are highly sensitive to hole size and shape deformation. In mid-2004, a purpose-built mPOF tower was commissioned that uses radiative heating (see Chapter 5). Fibre produced on this tower has generally been of greater consistency along its length. As a result, it was found that most fibres drawn on this tower had a reduced transmission loss (by around 50%) compared to fibre drawn from identical preforms

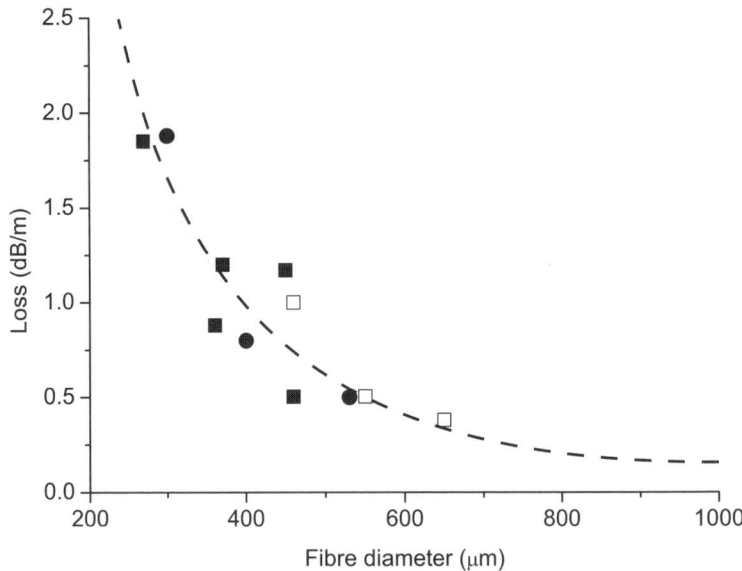

Fig. 7.11. Loss with respect to outer fibre diameter for different mPOF designs. After Large et al. [2006].

on a previous tower that employed convective heating. This is attributed to the much better structural uniformity (both longitudinally and transversely) and a reduced surface roughness.

Combining all the previous insights, an mPOF was fabricated with the lowest loss figure to date: 192 dB/km at 650 nm. The loss measurement is shown in Fig. 7.12 [Lwin et al. 2005a,b]. This figure is comparable to conventional PMMA-based POF. The fibre loss at short wavelengths is partially attributed to scattering (as indicated by the λ^{-4} dependence of the loss curve) originating from surface roughness within the holes in the fibre (possibly a residual artefact of the drilling process) and/or from residual micro-particles on the hole surfaces. Some of the remaining loss may arise from confinement loss and from the ingress of chemical impurities from the cutting fluid that are not flushed out by water. Overall though it is believed that mPOF transmission performance is now limited primarily by microbending [Lwin et al. 2005b] and confinement loss (see Section 3.2 and [White et al. 2001]).

Figure 7.13 shows a halving of the loss every 6 months since their first fabrication.

Clearly the use of a base material with a higher transparency would enable a significant further loss reduction. MPOFs can in principle achieve a loss much below the material loss, as is demonstrated with the hollow-core bandgap fibres in Chapter 8. Even in case of solid-core mPOF, a loss below the material loss should be obtainable since a fraction of the power of the

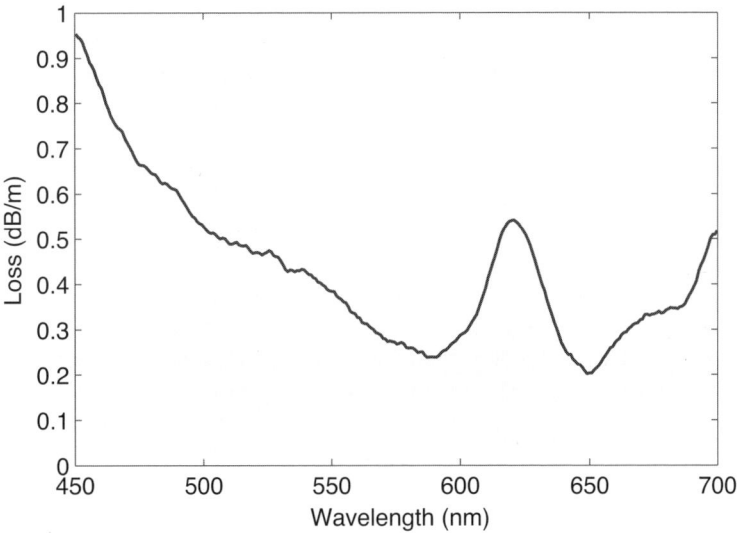

Fig. 7.12. Loss spectrum of a low-loss mPOF. After Lwin et al. [2005b] and Large et al. [2006].

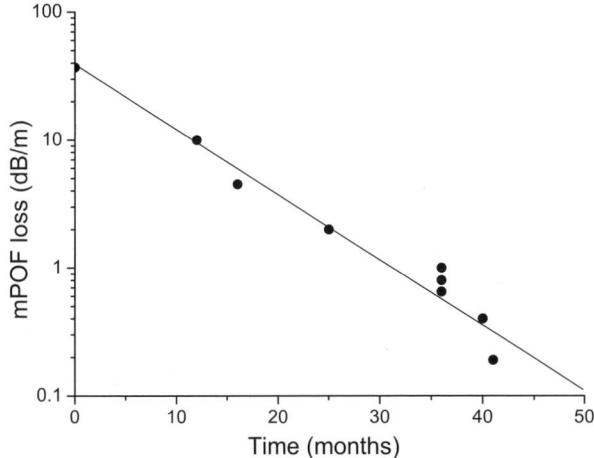

Fig. 7.13. Lowest recorded mPOF loss as a function of time starting from their first fabrication. The drawn line is a fit curve showing a halving of the loss every 6 months down to the material loss limit. After Large et al. [2006].

guided mode(s) is located in the air holes, and this fraction can be made as large as 50% [Monro et al. 2001].

7.2.6 Bandwidth

Optical fibres have clearly revolutionised long-haul telecommunications. However, for shorter distances there are a number of important applications that remain largely electronic, such as local area networks (LAN), fibre to the home (FTTH) and even chip-to-chip communication and interconnects. At these short distances, the attenuation of the fibre is less important than issues such as how easily the fibre can be connected, which is directly related to the size of the core. Single-mode glass fibres, for instance, have a very high bandwidth but are difficult to join because of their small core size. Fibres with much larger cores are difficult to make in glass, however polymer fibres can easily be made with cores of order 1 mm in diameter whilst still being flexible. For this reason many people believe that short-haul, high-speed connections will be made using polymer optical fibres (see also Section 1.5.3).

Large-core fibres however are multimode and suffer badly from modal dispersion which limits their bandwidth. This effect can be greatly reduced by using the appropriate refractive index profile across the fibre. A nearly parabolic profile was determined as ideal for conventional polymer fibre [Monroy et al. 2003] but it remains difficult to control the refractive index profile accurately enough, and the resulting fibres are relatively expensive. Potentially a specific type of mPOFs may offer a cheaper and easier solution to this long standing problem (see Chapter 9).

Bandwidth can be defined in both the frequency and time domains, and similarly the experimental characterisation concerns either the frequency response or the temporal behaviour of light pulses as they propagate down the fibre. Measuring bandwidth in the frequency domain involves scanning the modulation frequency of a light source over a range of frequencies and comparing this to the transmission through the fibre. The bandwidth is defined as the frequency where the transmitted signal is reduced by 3 dB [Kolesar and Mazzarese 2002]. Time domain methods are generally preferred since they can reveal more detailed information about pulse propagation and mode behaviour. The two main types of time-domain measurements are pulse width distortion (PWD) and differential mode delay (DMD) methods.

The pulse width distortion (PWD) method involves the launching of a pulse of known temporal width into an optical fibre, and examining the pulse spread over a length L of fibre [Okamoto 1979]. Light of a numerical aperture $NA \approx 0.1$ is launched at normal incidence into the end of a fibre. This NA is used since it is representative of a typical light source for a POF [Bachmann et al. 2001]. By comparing the width of the input pulse (σ_{input}) to the output pulse (σ_{output}) the timing delay (δt) and therefore the optical bandwidth (f), can be evaluated as

$$\delta t = \sqrt{\sigma_{\text{output}}^2 - \sigma_{\text{input}}^2}, \tag{7.1}$$

$$f = \frac{1}{2\delta t} \tag{7.2}$$

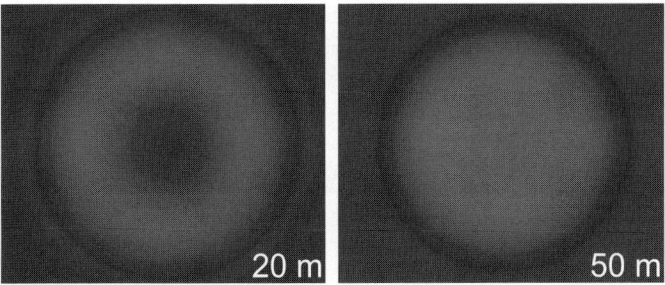

Fig. 7.14. Output fields of a Mitsubishi GK40 PMMA optical fibre taken at 20 m and 50 m lengths, using a 20° launch angle. The change in output from a ring to a disc indicates that the equilibrium length of this fibre is 20 m $< L_e <$ 50 m. *Images courtesy of Hans Poisel, Polymer Fiber Application Center (POFAC), Germany.*

The differential mode delay (DMD) method is a refinement of the PWD method [Bachmann et al. 2001]. The main difference is the use of a smaller launch $NA \approx 0.015$, so that only a small group of modes is excited. Timing delays between different mode groups are then examined and the bandwidth determined. A particular mode group can be selected by either varying the launch angle or the launch position in the core, the latter being generally used for microstructured fibres in order to avoid launching into holes, which can happen in translation. Given an output pulse signal $s(t)$, the central pulse time t_c can be evaluated [Bachmann et al. 2001] using

$$t_c = \frac{1}{A} \int_{-\infty}^{+\infty} t \cdot s(t) dt \qquad (7.3)$$

where A is the area below the curve $s(t)$. Using knowledge of the fibre NA, the difference between the minimum and maximum pulse propagation times within this launch angle range can be used to determine the bandwidth.

It should be noted that DMD measurements are only possible when the equilibrium length is longer than the length of fibre under test, ie. when the mode mixing is weak. This is generally the case for silica multimode fibres [Kolesar and Mazzarese 2002, Mortensen et al. 2003], but the equilibrium length for polymer multi-mode fibres is typically of the order of metres to tens of metres, and generally even shorter for mPOFs. Therefore an analysis of the bandwidth performance of an optical fibre requires a measurement of its equilibrium length, generally done by launching light into the fibre with a low NA at non-zero angle and examining the fibre output. As the light propagates down the fibre, modes exchange energy due to perturbations (see Section 2.4) causing the ring of light in the far-field output pattern to gradually become more fuzzy (Fig. 7.14). It eventually forms a filled-in disk [Savović and Djordjevich 2004]. This corresponds to a pseudo-steady state of power distribution over the modes and is used as a measure of the equilibrium length L_e [Simard et al. 2003]. This can also be measured using PWD for different lengths of fibre. If the pulse spread as a function of length exhibits a $\sqrt{L}$ dependence, then mode mixing is effective over that length [Shi et al. 1997, White et al. 1999].

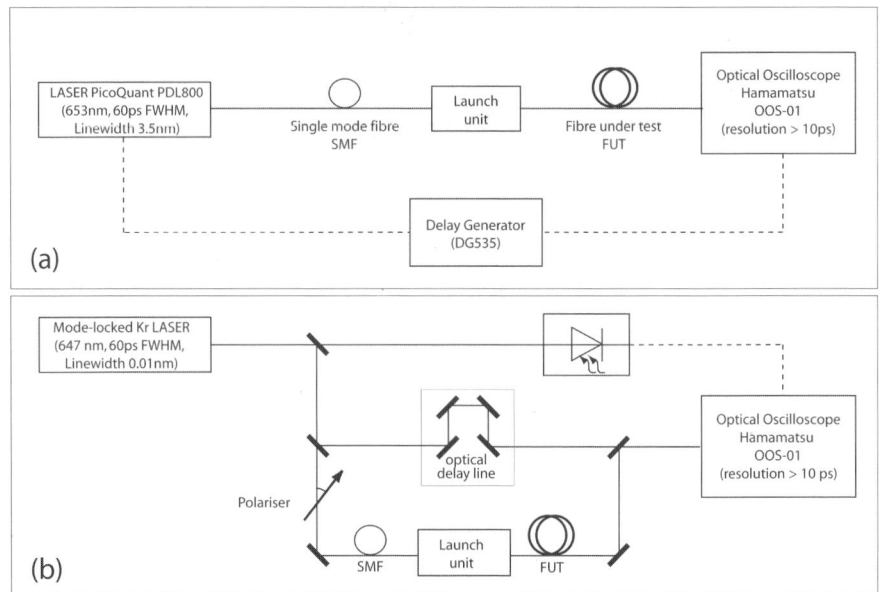

Fig. 7.15. Experimental setups used for multi-mode fibre bandwidth characterization. (a) An electronic delay generator is used to trigger the optical oscilloscope to the output pulse. (b) A free space optical delay line is used to synchronize the output pulse with the optical oscilloscope. *Experimental setup (a) designed by Alex Bachmann, Polymer Fiber Application Center (POFAC), Germany, and (b) designed by David Hirst, Southern Photonics, New Zealand.*

The specific experiment setup used for mPOF measurements by Bachmann et al. [2001] is shown in Fig. 7.15(a). A pulsed laser diode (PicoQuant PDL800) operating at $\lambda = 652$ nm with a spectral width of 3.5 nm couples a pulse with FWHM of 60 ps into a single mode fibre with NA = 0.1. There are two launch unit designs used to measure either DMD or PWD. For the DMD launch unit shown in Fig. 7.16(a), a fixed aperture and NA of 0.015 is used to launch onto the end-face of the fibre-under-test, which pivots about the focal point at variable angles. The PWD launch unit in Fig. 7.16(b) uses variable combinations of aperture diameters and microscope objectives to create a launch with an NA from 0.04 to 0.25, where the light is launched straight into the core. The output spot size from the single-mode fibre in both cased is $\approx 35\mu$m, and the temporal FWHM of 120 ps. A delay generator (Standford Research System DG535) is used to synchronize the delay from output pulse to the scan range of the optical oscilloscope. The optical oscilloscope (Hamamatsu OOS-01) has a temporal resolution better than 5 ps.

One of the issues affecting the experimental setup shown in Fig. 7.15(a) is the pulse broadening caused by the extent of electronic jitter from the delay generator. The influence of jitter on the output pulse width is dependent on the trigger level used on the delay generator. This is especially significant for the low trigger levels used to measure the reference input pulse. To overcome this problem, the electronic delay generator is replaced by a free space optical delay line, as illustrated in Fig. 7.15(b). In this setup, the quality of the reference input pulse is improved as it is measured directly by the optical oscilloscope. In addition, the input and output pulses are monitored on the optical oscilloscope simultaneously. The input laser used is a mode-locked Kr Laser at $\lambda = 647$ nm with a spectral width of 0.01 nm and a pulse width FWHM of 60 ps. The very small spectral width means that material dispersion can be neglected. Thus, comparing results from the two different setups allows the effect of chromatic dispersion to be isolated.

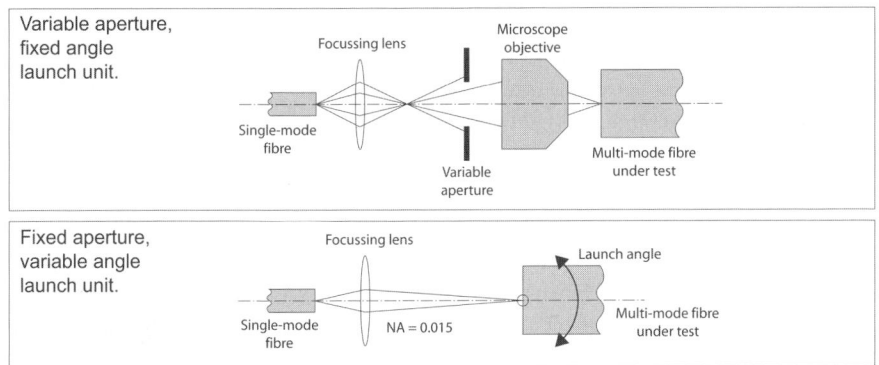

Fig. 7.16. The two launch units used for DMD and PWD measurements. *Experimental setup designed by Alex Bachmann, Polymer Fiber Application Center (POFAC), Germany.*

An example of a set of pulse profiles after transmission through an Optimedia fibre (OM GIGA 1000 μm) as recorded during a DMD measurement is shown in Fig. 7.17. A low *NA* of 0.015 at 652 nm was used to launch into a 50 m length of this fibre with a pulse width of 120 ps (FWHM). For clarity, the pulse amplitudes in Fig. 7.17 have been normalised and shifted vertically according to the launch angle, which was varied from -20 to +20°. This makes the *relative* delay between pulses transmitted in different modes evident.

All measurements of the bandwidth properties of GImPOFs are presented separately in Chapter 9.

The effects of dispersion-induced pulse broadening have also been measured on the transmission of THz pulses through a polymer photonic crystal

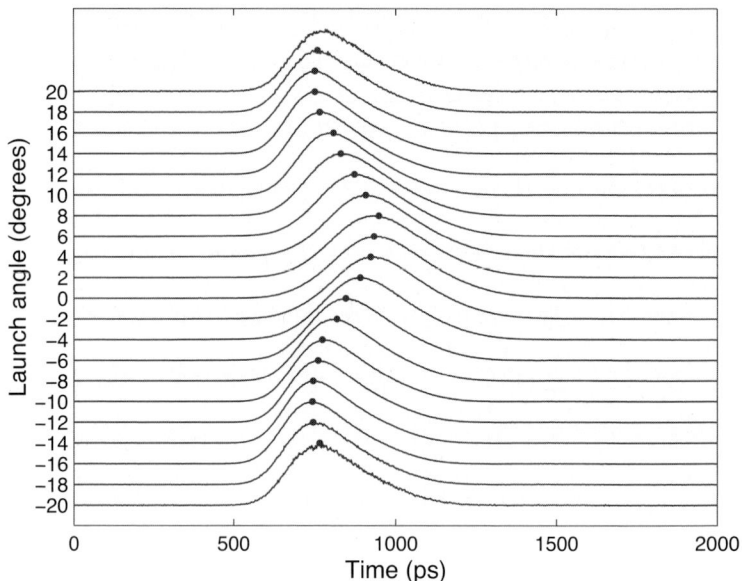

Fig. 7.17. Transmission pulses for a range of launch angles as recorded in a typical differential mode delay measurement. *Measurements performed by R. Lwin at the Polymer Fiber Application Center (POFAC), Germany.*

fibre made from stacked high-density polyethylene (HDPE) tubes of 500 µm diameter with a solid rod included as a central core [Han et al. 2002]. Guiding was observed from 0.1 to 3 THz, with loss determined by material absorption. The temporal shape of THz pulses transmitted through 2 cm of this structure was reported for radiation of 0.4 THz, and showed a pulse broadening from an input of 0.8 ps to an output pulse of 5 ps (FWHM) mainly as a result of the waveguide dispersion. This result was in good agreement with calculations and corresponds to a group velocity dispersion of 2 ps/THz/cm [Han et al. 2002].

7.2.7 Birefringence

Birefringence and the associated polarisation effects play an important role in a variety of applications and specialty fibre such as polarisation maintaining fibre [Noda et al. 1986, Rashleigh 1983]. This section outlines various methods to measure linear birefringence in mPOFs and illustrates this by applying the most commonly used method to determine the birefringence of an mPOF with uniformly oriented elliptical holes [Issa et al. 2004].

In a linearly birefringent fibre, two orthogonally linearly polarised modes are guided with slightly different propagation constants (see also Sections 2.1.1 and 4.4.5). If light is injected into the fibre so that both modes are excited, the relative phase difference of the two modes will grow as they propagate.

When this phase difference is an integral multiple of 2π, the input state is reproduced. Thus the effect of uniform birefringence is to cause a general polarisation state to evolve through a periodic sequence of states as it propagates. The length over which this beating occurs is the fibre beat length L_B. Commonly available (conventional) fibres have beat lengths of the order of centimetres. Microstructured optical fibres are capable of producing extremely large birefringence and can have a very short beat lengths of below a millimetre [Ortigosa-Blanch et al. 2000]. The basic methods available for measuring birefringence in optical fibres are (i) fibre cut-back, (ii) Rayleigh scattering, (iii) external modulation, and (iv) transmission-analyser methods.

The first method is performed by launching light into the fibre at 45° to the (suspected) birefringence axes and measuring the fibre output polarisation. The fibre is then cut back and the measurements repeated, revealing the beat length. This method however is destructive and limited in accuracy, especially for short beat lengths, though it is applicable to all wavelengths.

The second, most illustrative method to measure birefringence is to observe the Rayleigh-scattered light along a line perpendicular to the fibre axis and at 45° to both fibre modes. When both polarisation modes are equally excited, the scattered light is zero when the guided light is polarised along the direction of observation (45°). The distance between adjacent minima is equal to $L_B/2$ at that particular wavelength. This technique is more difficult to use for nonvisible wavelengths since Rayleigh scattering is weaker in the infrared, and (less sensitive) cameras or IR viewers must be employed.

In the third method, a local disturbance is created on the fibre which is moved along the propagation direction [Peyrilloux et al. 2003]. This disturbance can be a simple local lateral pressure on the fibre, or alternatively a precision (electro)magnet that locally disturbs the section of fibre due to the Faraday effect. The observed period of the output power of the fibre as the disturbance is translated corresponds to the beat length. This technique is limited by the need to concentrate the disturbance into a section of fibre that is significantly shorter than L_B, and in the case of the magneto-optic disturbance, the low Verdet constant of most fibre materials is a limiting factor. The lateral pressure modulation method and the Rayleigh scattering method were actually found inappropriate for microstructured fibres due to the effects of the structured cladding [Ortigosa-Blanch et al. 2000]. The magneto-optical modulation, however, was used successfully to measure relatively long (several cm) birefringence beat lengths in a MOF [Peyrilloux et al. 2003].

The fourth and more indirect transmission-analyser method has been frequently used due to its experimental simplicity and accuracy. Light is launched at 45° to the birefringence axes and a second polariser is positioned before the detector or spectrum analyser. In combination with a broadband source or a tunable laser this yields a transmission spectrum that shows (nearly) periodic features with wavelength that are related to the beat length. This method has been used successfully to measure birefringence of MOFs [Ortigosa-Blanch et al. 2000] and is demonstrated below for use with mPOFs.

Microstructured optical fibres with elliptical holes have received significant theoretical analysis because of their polarisation dependent interaction with guided electromagnetic waves. In addition to the high-birefringence and polarisation maintaining characteristics of such fibres [Steel and Osgood 2001, Mogilevtsev et al. 2001], a number of properties such as photonic bandgap guidance [Qiu and He 1999] and dispersion [Steel and Osgood 2001] have been investigated. The addition of an elliptical hole in the core region of MOFs has also been studied [Mogilevtsev et al. 2001, Zhi et al. 2003] as a means of enhancing birefringence even beyond the exceptionally high levels demonstrated in MOFs to date [Ortigosa-Blanch et al. 2000, Suzuki et al. 2001]. Birefringence measurements on an mPOF with uniformly oriented elliptical holes is outlined here, providing the first steps at experimentally realising the variety of elliptical hole structures presented in the literature [Issa et al. 2004].

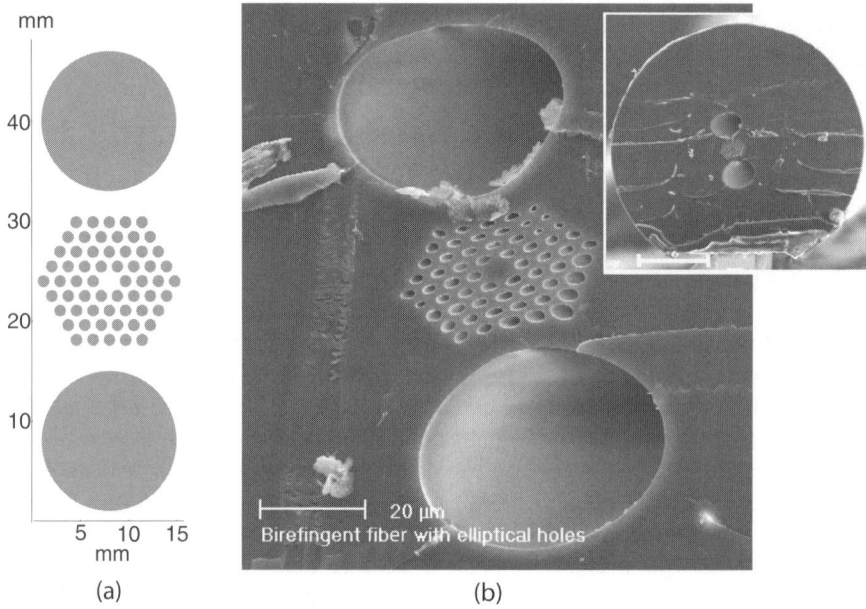

(a)   (b)

Fig. 7.18. (a) Schematic showing the drilled hole pattern in a birefringent PMMA preform. (b) Scanning electron microscope image of the resulting sleeved fibre after drawing. After Issa et al. [2004].

While hole deformation effects during fibre drawing are often unwelcome, they can be advantageously employed to create oriented elliptical holes, given the appropriate preform geometry. The preform hole pattern shown in Fig. 7.18(a) was fabricated and drawn to fibre of three different outer diameters. The deformed hole pattern in the resulting fibre is shown in the SEM image of Fig. 7.18(b).

A close-up of the fibre structure is shown in Fig. 7.19 with the parameters $\Lambda_a, \Lambda_b, d_b$ and d_a indicated. The major hole pitch, Λ_a, for the three different diameter fibres are 5.12, 3.82 and 3.14 µm. The ellipticity of the cores, defined as the ratio of minor to major pitches, Λ_b/Λ_a, are 0.61, 0.61 and 0.59, respectively. The average hole major diameters, d_a, of the inner ring are 3.0, 2.1 and 1.6 µm for these three fibres, with average ellipticity d_b/d_a of 0.59, 0.54 and 0.48, respectively. From the reproducibility of these measurements retaken over several perpendicular cleaves, the accuracy is estimated to be ±5%.

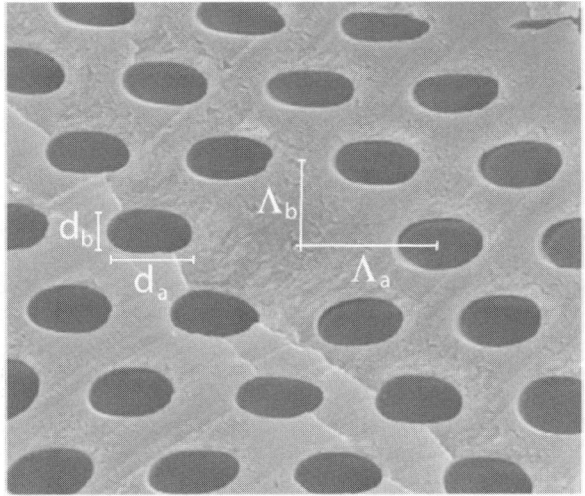

Fig. 7.19. (a) SEM image of inner rings of elliptical holes and definition of structural parameters. After Issa et al. [2004].

Polarised light from a broadband light source was launched at 45° to the principal axes of the birefringent mPOF. The output polarisation state was analysed with a second polariser also at 45° to the principal axis. An intensity modulation was observed over a broad wavelength range (see Fig. 7.20) which could be used to characterise the wavelength dependent modal birefringence, $B(\lambda) = n_x - n_y = \lambda/L_B(\lambda)$ where n_x, n_y are the effective indices of the dominantly polarised modes and L_B is the polarisation beat length. The light source used for these measurements was a high brightness white light source. However, for fibres with very short modulation periods, a tunable Ti:Sapphire laser operating from 700 to 860 nm was used. Measurements at longer wavelengths were impossible due to very high PMMA material absorption, limiting the measurements to the transparency windows in the visible. All fibres were experimentally found to be single-mode. However, the fibre with $\Lambda_a = 5.12$ µm was only single-mode for lengths over two metres.

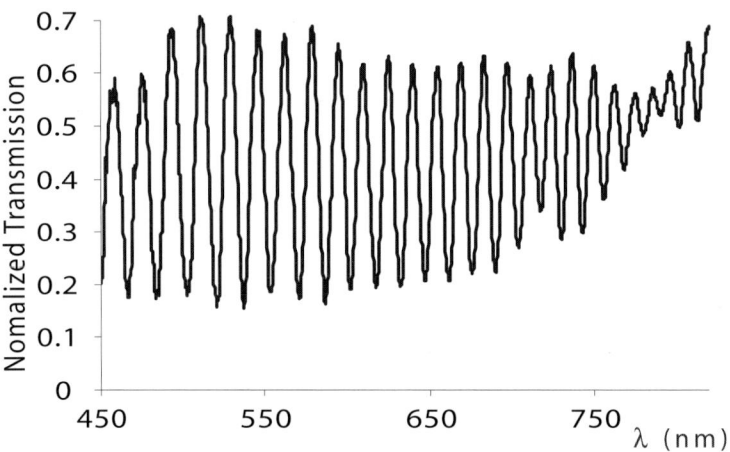

Fig. 7.20. Example transmission-analyser spectrum for birefringence measurements. The effects of a PMMA absorption peak near 770 nm are observable. After Issa et al. [2004].

The relative phase difference, ϕ, between the two polarisation modes after a length L is given by $\phi(\lambda) = 2\pi LB/\lambda$. Making the assumption that $B \propto (\lambda/\Lambda_a)^{k_0}$ [Ortigosa-Blanch et al. 2000, Suzuki et al. 2001], the use of $\Delta\phi = \phi(\lambda(1 + \Delta\lambda/\lambda)) - \phi(\lambda)$ gives

$$B(\lambda) = \frac{\Delta\phi\ \lambda}{2\pi L} \left[\left(1 + \frac{\Delta\lambda}{\lambda}\right)^{k_0 - 1} - 1 \right]^{-1} \qquad (7.4)$$

which is accurate for large $\Delta\lambda/\lambda$. When λ is centred on a peak and $\Delta\lambda$ represents the distance to a nearby peak, then $\Delta\phi = 2\pi n$ where n is an integer. Alternatively, a peak-to-node reading corresponds to $\Delta\phi = 2\pi(n + 1/2)$. In this way, several measurements can be averaged at each sample wavelength to statistically reduce the overall errors. The unknown exponent, k_0, can be determined efficiently by further curve-fitting methods.

The resulting birefringence measurements on the three fibres are overlayed in Fig. 7.21, where the wavelength has been scaled against the major pitch of the fibres. In good accordance with the size-wavelength scaling law for electromagnetism [Sakoda 2001], the three measurements piecewise construct a distinct functional dependence of birefringence on wavelength. The small discontinuities between measurements are the result of the slightly different deformations at different microstructure sizes and are approximately the same magnitude as the $\pm 5\%$ errors in the data. The measured values of $k_0 \simeq 2.8$ are in excellent agreement between fibres and support the appropriateness of a power-law fit over a very broad wavelength range.

Numerical modelling of this mPOF structure was also carried out using a highly efficient, fully vectorial mode solver [Issa and Poladian 2003] employ-

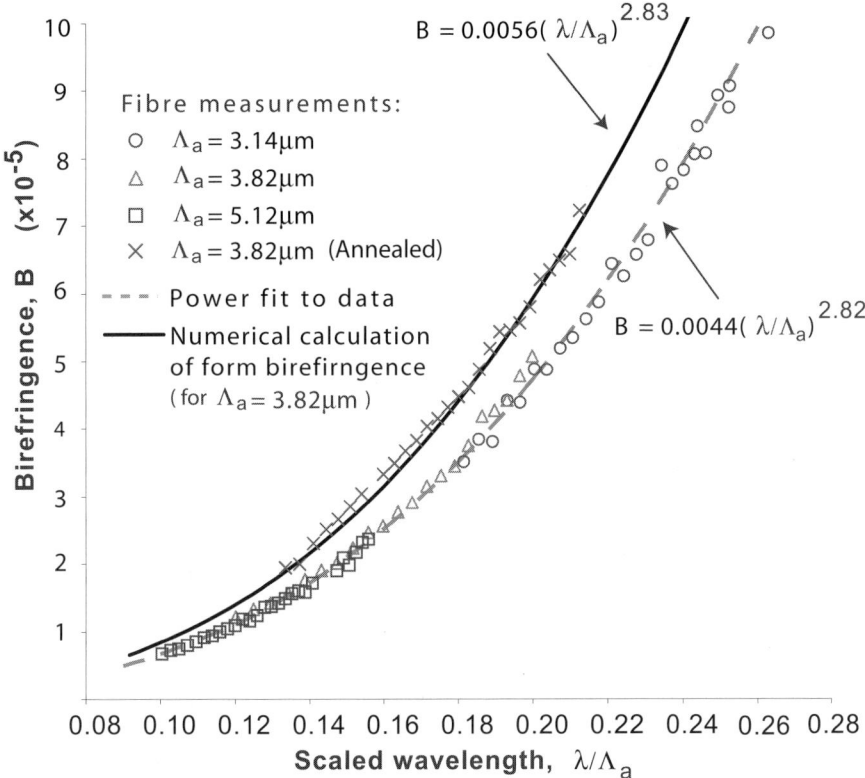

Fig. 7.21. Measured and numerically calculated birefringence. The stated power-law for the numerical calculation is obtained from a fit with $R^2 > 0.999$. After Issa et al. [2004].

ing adjustable boundary conditions, using a finite difference implementation in the radial direction and an azimuthal Fourier decomposition. The confinement loss of the fundamental modes was found to be negligible in comparison to the material loss. Only waveguide-induced (form) birefringence was calculated with the stress-optic contribution neglected. Most importantly, the form birefringence was found to be weakly dependent on the measured deformations.

The numerical results for the mPOF with $\Lambda_a = 3.82$ μm are plotted in Fig. 7.21 and show poor agreement with the initial measurements. Subsequently, this fibre was annealed at 90°C (~ 25°C below the glass transition temperature) for 25 min to alleviate any material stresses and the birefringence measurements retaken (shown in Fig. 7.21). The annealed results show excellent agreement with numerical modelling and demonstrate that a non-zero stress-optic birefringence was latent in the material in opposition to the form birefringence. The experimental results indicate that the increase in measured birefringence by eliminating stress was ∼ 25% which is wavelength independent. The birefringence for these fibres increases strongly with wavelength and beyond 850 nm is comparable in magnitude with conventional polarisation maintaining fibres, for which birefringence is nearly wavelength independent. SEM images of the annealed fibre confirmed no discernible change in the microstructure and thus the purely form birefringence of the type investigated here is expected to be largely independent of temperature [Szpulak et al. 2002].

References

Achenbach, C P and Cobb, J H (2003). Computational studies of light acceptance and propagation in straight and curved multimode active fibers. *Journal of Optics A*, 5:239–49.

Asnaghi, D, Gambirasio, A, Macchetta, A, Sarchi, D, and Tassone, F (2002). Fabrication of a large effective-area microstructured plastic optical fibre: design and transmission tests. In *Proceedings of the European Conference on Optical Communications*, volume 3, pages 632–3, Copenhagen, Denmark.

Bachmann, A, Klein, K F, Poisel, H, Ziemann, O, and Niewisch, J (2001). Differential mode delay (DMD) measurements on graded index POF. In *Proceedings of the International Plastic Optical Fibres conference*, volume 10, pages 57–61, Amsterdam, the Netherlands.

Barton, G, van Eijkelenborg, M A, Henry, G, Issa, N A, Klein, K-F, Large, M C J, Manos, S, Padden, W, Pok, W, and Poladian, L (2003). Characteristics of multimode microstructured pof performance. In *Proceedings of the International Plastic Optical Fibres conference*, volume 12, pages 81–84, Seattle, USA.

Bouwmans, G, Percival, R M, Wadsworth, W J, Knight, J C, and Russell, P St J (2003). High-power Er:Yb fiber laser with very high numerical aperture pump-cladding waveguide. *Applied Physics Letters*, 83:817–8.

Bucknall, C B (1977). *Toughened plastics*. Applied Science Publications, London, UK.

Canning, J, Buckley, J E, Groothoff, N, Davies, B-L, and Zagari, J (2002). Uv laser cleaving of air-polymer structured fibre. *Optics Communications*, 202(1-3):139–43.

Choi, J, Cha, H, and Lee, J (2003). Single mode polymer photonic crystal fibers. In *Proceedings of the International Plastic Optical Fibres conference*, volume 12, University of Washington, Seattle, USA.

Choi, J, Kim, D Y, and Paek, U C (2001). Fabrication and properties of polymer photonic crystal fibers. In *Proceedings of the International Plastic Optical Fibres Conference*, pages 335–60, Amsterdam, Netherlands.

Daum, W, Krauser, J, Zamzow, P E, and Ziemann, O (2002). *POF Polymer Optical Fibers for Data Communication*. Springer Verlag, Berlin, Germany, first edition.

Franzen, D L, Young, M, Cherin, A H, Head, E D, Hackert, M, Raine, K, and Baines, J (1989). Numerical aperture of multimode fibers by several methods: Resolving differences. *Journal of Lightwave Technology*, 7(6):896–901.

Gordon, J E (1991). *The new science of strong materials, or, Why you don't fall through the floor*. Penguin, London, UK, second edition.

Goto, M, Quema, A, Takahashi, H, Ono, S, and Sarukura, N (2004). Teflon photonic crystal fiber as terahertz waveguide. *Japanese Journal of Applied Physics*, 43:L317–9.

Han, H, Park, H, Cho, M, and Kim, J (2002). Terahertz propagation in a plastic photonic crystal fiber. *Applied Physics Letters*, 80(15):2634–6.

Huang, C, Ho, M, Cheng, C, Ma, K, Kiang, Y, Chang, H, and Yang, C C (2004a). Design, fabrication, and characterization of polymer microstructured fiber. In *Proceedings of the Photonics West Conference*, San Jose, CA, USA.

Huang, C W, Ho, M-C, Chien, H H, Ma, K J, Zheng, Z P, Yu, C P, Chang, H C, and Yang, C C (2003). Design, fabrication, and characterization of microstructured polymer optical fibers. In *Proceedings of the Conference on Lasers and Electro Optics Pacific-Rim*, volume 1, page 18, Taipei, Taiwan.

Huang, C W, Ho, M C, Yu, C P, Chang, H C, Yang, C C, Chien, H H, Ma, K J, and Zheng, Z P (2004b). Fabrication and characterization of microstructured polymer optical fibres. In *Proceedings of the Conference on Lasers and Electro Optics*, page CThX2, San Francisco, USA.

Issa, N A (2004). High numerical aperture in multimode microstructured optical fibers. *Applied Optics*, 43(33):6191–7.

Issa, N A and Padden, W E (2004). Light acceptance properties of multimode microstructured optical fibers: Impact of multiple layers. 12(14).

Issa, N A and Poladian, L (2003). Vector wave expansion method for leaky modes of microstructured optical fibres. *Journal of Lightwave Technology*, 21(4):1005–12.

Issa, N A, van Eijkelenborg, M A, Fellew, M, Cox, F, Henry, G, and Large, M C J (2004). Fabrication and study of microstructured optical fibers with elliptical holes. *Optics Letters*, 29(12):1336–8.

Issa, N A, von Korff Schmising, C, Padden, W E, and van Eijkelenborg, M A (2005). High numerical aperture in large-core microstructured optical fibres. In *Proceedings of the SPIE Photonics West conference*, pages 5691–13, San Jose, CA, USA.

Kausch, H H (1978). *Polymer fracture*. Springer-Verlag, New York, USA.

Knight, J C, Birks, T A, and Russell, P St J (2001). "Holey" silica fibers. In Markel, V A and George, T F, editors, *Optics of Nanostructured Materials*, chapter 2, pages 39–71. Wiley, New York, USA.

Knight, J C, Birks, T A, Russell, P St J, and Atkin, D M (1996). All-silica single mode optical fiber with photonic crystal cladding. *Optics Letters*, 21(19):1547–9.

Kolesar, P F and Mazzarese, D (2002). Understanding multimode bandwidth and differential mode delay measurements and their applications. In *Proceedings of the International Wire and Cable Symposium of IWCS Inc.*, volume 51, pages 453–60, Lake Buena Vista, Florida, USA.

Kondo, S, Ishigure, T, and Koike, Y (2004). Fabrication of polymer photonic crystal fiber. In *Proceedings of the Micro-Optics Conference*, volume 10, pages B–7, Jena, Germany.

Kuhlmey, B T, McPhedran, R C, and de Sterke, C M (2002). Modal cutoff in microstructured optical fibers. *Optics Letters*, 27(19):1684–6.

Kusy, R P (1977). The ductile-brittle transition of acrylic. *Journal of Non-Crystalline Solids*, 24(1):141–4.

Kusy, R P and Turner, D T (1977). Influence of the molecular weight of poly(methyl methacrylate) on fracture morphology in notched tension. *Polymer*, 18(4):391–402.

Kuzyk, M G (2003). An overview of dye-doped polymer optical fibers: Fabrication characterization and applications. In *Proceedings of the International Plastic Optical Fibres conference*, volume 12, pages 73–6, Seattle, Washington, USA.

Large, M C J, Ponrathnam, S, Argyros, A, Bassett, I, Punjari, N S, Cox, F, Barton, G W, and van Eijkelenborg, M A (2006). Microstructured polymer optical fibres: New opportunities and challenges. In Burillo, G, Ogawa, T, Rau, I, and Kajzar, F, editors, *Molecular Crystals and Liquid Crystals Journal, Special issue, Proceedings of the 8th international conference on frontiers of polymers and advanced materials*, volume 446, pages 219–31. Taylor & Francis.

Law, S H, Harvey, J D, Kruhlak, R J, Song, M, Wu, E, Barton, G W, van Eijkelenborg, M A, and Large, M C J (2006a). Cleaving of microstructured polymer optical fibres. *Optics Communications*, 258(2):193–202.

Law, S H, Lwin, R, Gan, J, van Eijkelenborg, M A, Barton, G W, and Yan, C (2006b). Cleaved end-face quality of microstructured polymer optical fibres. *Optics Communications*, 265:513–20.

Lwin, R, Barton, G, Keawfanapadol, T, Large, M, Poladian, L, Tanner, R, van Eijkelenborg, M A, and Xue, S (2005a). Suspended core microstructured polymer optical fibre: Connecting to reality. In *Proceedings of the Australian Conference on Optical Fibre Technology*, volume 30, Star City, Sydney, Australia.

Lwin, R, Barton, G, Large, M C J, Poladian, L, and van Eijkelenborg, M A (2005b). Progress on low loss of microstructured polymer optical fibres. In *Proceedings of the International Plastic Optical Fibres conference*, volume 14, pages 37–40, Hong Kong, China.

Mark, J E (2004). *Physical properties of polymers*. Cambridge University Press, Cambridge, UK, third edition.

Matsushige, K, Radcliffe, S V, and Baer, E (1976). The pressure and temperature effects on brittle-to-ductile transition in PS and PMMA. *Journal of Applied Polymer Science*, 20(7):1853–66.

McCrum, N G, Buckley, C P, and Bucknall, C B (1997). *Principles of polymer engineering*. Oxford University Press, Oxford, UK, second edition.

Mediavilla, E, Arribas, S, and Watson, F (1998). *Fiber optics in astronomy Vol 3*. Astronomical Society of the Pacific, San Francisco, USA.

Mogilevtsev, D, Broeng, J, Barkou, S E, and Bjarklev, A (2001). Design of polarization-preserving photonic crystal fibers with elliptical pores. *Journal of Optics A*, 3:S141–3.

Monro, T M, Belardi, W, Furusawa, K, Baggett, J C, Broderick, N G R, and Richardson, D J (2001). Sensing with microstructured optical fibres. *Measurement Science Technology*, 12:854–8.

Monroy, I Tafur, vd Boom, H P A, Koonen, A M J, Khoe, G D, Watanabe, Y, Koike, Y, and Ishigure, T (2003). Data transmission over polymer optical fibers. *Optical Fiber Technology*, 9:159–71.

Mortensen, N A and Folkenberg, J R (2002). Near-field to far-field transition of photonic crystal fibers: symmetries and interference phenomena. *Optics Express*, 10(11):475–81.

Mortensen, N A, Stach, M, Broeng, J, Petersson, A, Simonsen, H R, and Michalzik, R (2003). Multi-mode photonic crystal fibers for VCSEL based data transmission. *Optics Express*, 11(17):1953–9.

Nielsen, M D, Vienne, G, Folkenberg, J R, and Bjarklev, A (2003). Investigation of microdeformation-induced attenuation spectra in a photonic crystal fiber. *Optics Letters*, 28(4):236–8.

Noda, J, Okamoto, K, and Sasaki, Y (1986). Polarization-maintaining fibers and their applications. *Journal of Lightwave Technology*, 4(8):1071–89.

Okamoto, K (1979). Comparison of calculated and measured impulse responses of optical fibers. *Applied Optics Letters*, 18(13):2199–206.

Onishi, T (2001). Low loss perfluorinated GI-POF. In *Proceedings of the International Plastic Optical Fibres conference*, pages 337–40, Amsterdam, The Netherlands.

Ortigosa-Blanch, A, Knight, J C, Wadsworth, W J, Arriaga, J, Mangan, B J, Birks, T A, and Russell, P St J (2000). Highly birefringent photonic crystal fibers. *Optics Letters*, 25(18):1325–27.

Park, J H, Shin, B G, and Kim, J J (2002). Fabrication of plastic holey fibers. In *Proceedings of the International Plastic Optical Fibres conference*, volume 11, pages PD9–11, Tokyo, Japan.

Passaglia, E (1987). Crazes and fracture in polymers. *Journal of physics and chemistry of solids*, 48(11):1075–100.

Peyrilloux, A, Chartier, T, Hideur, A, Berthelot, L, Mélin, G, Lempereur, S, Pagnoux, D, and Roy, P (2003). Theoretical and experimental study of the birefringence of a photonic crystal fiber. *Journal of Lightwave Technology*, 21(2):536–9.

Qiu, M and He, S (1999). Large complete band gap in two dimensional photonic crystals with elliptical air holes. *Physical Review B*, 60:10610–12.

Rashleigh, S C (1983). Origins and control of polarisation effects in single-mode fibers. *Journal of Lightwave Technology*, 1(2):312–31.

Rennsteig (2005). Tools for cutting polymeric optical fibres.

Sahu, J K, Renaud, C C, Furusawa, K, Selvas, R, Alvarez-Chavez, J A, Richardson, D J, and Nilsson, J (2001). Jacketed air-clad cladding pumped ytterbium-doped fiber laser with wide tuning range. *Electronics Letters*, 37(18):1116–7.

Sakoda, K (2001). *Optical properties of photonic crystals*. Springer, Berlin, Germany.

Savović, S and Djordjevich, A (2004). Influence of numerical aperture on mode coupling in step-index plastic optical fibres. *Applied Optics*, 43(29):5542–6.

Shi, R F, Koeppen, C, Jiang, G, Wang, J, and Garito, A F (1997). Origin of high bandwidth performance of graded-index plastic optical fibers. *Applied Physics Letters*, 71(25):3625–67.

Shin, B-G, Park, J-H, and Kim, J-J (2004). Plastic photonic crystal fiber fabricated by centrifugal deposition method. *Journal of Nonlinear Optical Physics and Materials*, 13(3-4):519–23.

Simard, M, Carlson, M, Babin, F, and Trembly, M (2003). Understanding launch conditions for multimode connector and cable assembly testing. *Wave Review*, 9(3).

Steel, M and Osgood, R M (2001). Polarization and dispersive properties of elliptical-hole photonic crystal fibers. *Journal of Lightwave Technology*, 19(4):495–503.

Suzuki, K, Kubota, H, Kawanishi, S, Tanaka, M, and Fujita, M (2001). Optical properties of low-loss polarization-maintaining photonic crystal fiber. *Optics Express*, 9(13):676–80.

Szpulak, M, Martynkien, T, Urbańczyk, W, Wójcik, J, and Bock, W J (2002). Influence of temperature on birefringence and polarization mode dispersion in photonic crystal holey fiber. In *Proceedings of the 1st European Symposium on Photonic Crystals*, volume 2, pages 89–92, Warsaw, Poland.

Tunnermann, A, Schreiber, T, Röser, F, Liem, A, Höfer, S, Zellmer, H, Nolte, S, and Limpert, J (2005). The renaissance and bright future of fibre lasers. *Journal of Physics B*, 38:S681–93.

van Eijkelenborg, M A, Argyros, A, Bachmann, A, Barton, G W, Large, M C J, Henry, G, Issa, N A, Klein, K F, Poisel, H, Pok, W, Poladian, L, Manos, S, and Zagari, J (2004). Bandwidth and loss measurements of graded-index microstructured polymer optical fibre. *Electronics Letters*, 40(10):592–3.

van Eijkelenborg, M A, Issa, N A, von Korff-Schmising, C, and Hiscocks, M (2006). Rectangular-core microstructured polymer optical fibers for interconnect applications. *Electronics Letters*, 42(4):201–2.

van Eijkelenborg, M A, Large, M C J, Argyros, A, Zagari, J, Manos, S, Issa, N A, Bassett, I, Fleming, S, McPhedran, R C, de Sterke, C M, and

Nicorovici, N A P (2001). Microstructured polymer optical fibre. *Optics Express*, 9(7):319–27.

Wadsworth, W J, Joly, N, Knight, J C, Birks, T A, Biancalana, F, and Russell, P St J (2004a). Supercontinuum and four-wave mixing with Q-switched pulses in endlessly single-mode photonic crystal fibers. *Optics Express*, 12(2):299–309.

Wadsworth, W J, Percival, R M, Bouwmans, G, Knight, J C, Birks, T A, Hedley, T D, and Russell, P St J (2004b). Very high numerical aperture fibers. *IEEE Photonics Technology Letters*, 16(3):843–5.

White, T P, McPhedran, R C, de Sterke, C M, Botten, L C, and Steel, M J (2001). Confinement losses in microstructured optical fibers. *Optics Letters*, 26(21):1660–2.

White, W R, Dueser, M, Reed, W A, and Onishi, T (1999). Intermodal dispersion and mode coupling in perfluorinated graded-index plastic optical fiber. *IEEE Photonics Technology Letters*, 11(8):997–9.

Zhi, W, Guobin, R, Shuqin, L, and Shuisheng, J (2003). Dependence of mode characteristics on the central defect in elliptical hole photonic crystal fibers. *Optics Express*, 11(17):1966–79.

8

Hollow-Core Microstructured
Polymer Optical Fibres

When a book and a head collide and there is a hollow sound, is it always from the book?

Georg Christoph Lichtenberg (1742 - 1799)

This chapter is devoted to mPOFs that guide in a hollow core. It describes two methods for fabricating these fibres and explains their transmission characteristics. Common issues affecting the performance are discussed as well as the various applications.

8.1 Introduction

The ability to guide light in a hollow core is of particular relevance to polymer optical fibres as the material absorption of polymers is an important limitation. Air-guidance allows such absorption to be largely overcome, allowing for example, the transmission of light in the infrared with a loss well below that of the material [Argyros et al. 2006c]. Guidance in a hollow core reduces the effects of material absorption and nonlinearity allowing high-power beam delivery even at wavelengths where the material's transmission is very low [Temelkuran et al. 2002, Humbert et al. 2004, Shephard et al. 2005].

Although conventional POF is inexpensive both in terms of the material and processing costs, the high material absorption restricts the wavelengths and distances over which it is usable. Figure 8.1 shows the absorption spectrum of PMMA, extending to a wavelength of 1650 nm. Transparency can be improved by using deuterated or fluorinated polymers such as CYTOP® [Koike 1996], but these are very expensive for use in high volumes (see also Section 1.1).

Thus, while polymer is a more attractive material to work with than silica — it is cheaper, easier to process, more flexible and lighter — it has always been constrained as an optical material by its high absorption. Photonic bandgap guidance in a hollow core offers a possible solution to this as

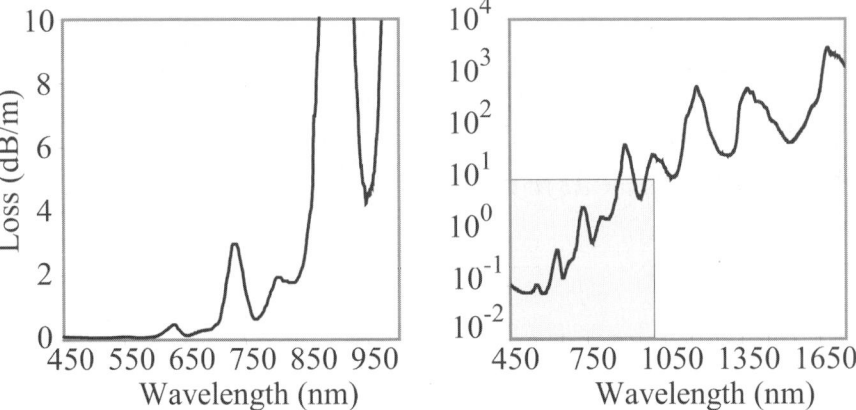

Fig. 8.1. Material absorption spectrum of commercially available high-grade PMMA. Note that the logarithmic scale on the right graph covers 6 orders of magnitude. The gray region corresponds to the left graph which has a linear scale. After Argyros et al. [2006c].

the fibre properties are predominantly determined by air and the microstructure rather than the polymer material properties. For an explanation of the bandgap guidance mechanism please see Section 3.1.

8.2 Fabrication

The first hollow-core mPOF (HC-mPOF) was made using a drilled preform as shown in Fig. 8.2(a). The cladding holes were pressurised during the cane draw whilst the central hole was kept at atmospheric pressure. This was achieved by modifying the extender used to hold the preform in the furnace, as shown in Fig. 8.2(b). The high pressure served to increase the air fraction in each ring of holes and to partially collapse the core, as shown previously in Fig. 6.10. The cane was then sleeved and drawn under normal conditions which allowed the central hole to (re-)expand back to the desired size and stretch the surrounding microstructure. Examples of fibres produced using this method are shown in Fig. 8.3.

The capillary-stacking method, described in Sec. 5.1.2, has also been used to make HC-mPOF. The preform shown in Fig. 5.4 was drawn to a cane of 12 mm diameter (no internal pressure applied) with the longer preform length allowing for a "steady state" draw to be achieved. This latter point was reflected in the quantity, quality and consistency of the cane along the length produced, the cross-section of which is shown in Fig. 8.4. This 12 mm cane was redrawn to a 6 mm diameter for sleeving. During the fibre draw, the

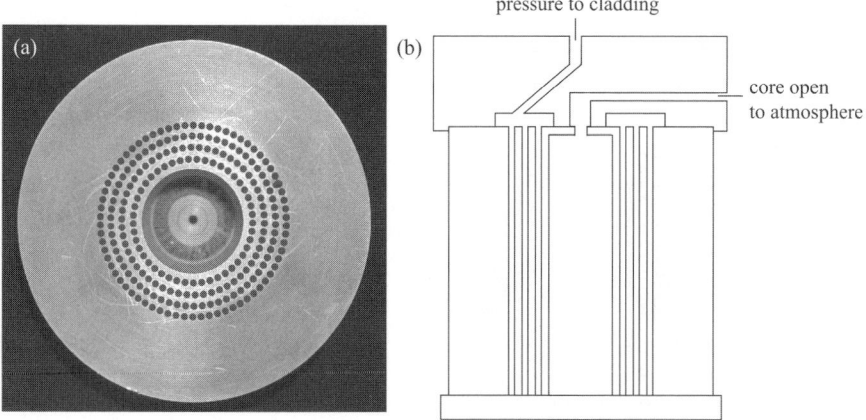

Fig. 8.2. (a) A drilled preform viewed from below with the central hole only partially open at the top. (b) Design of the extender used to apply pressure to the cladding holes only.

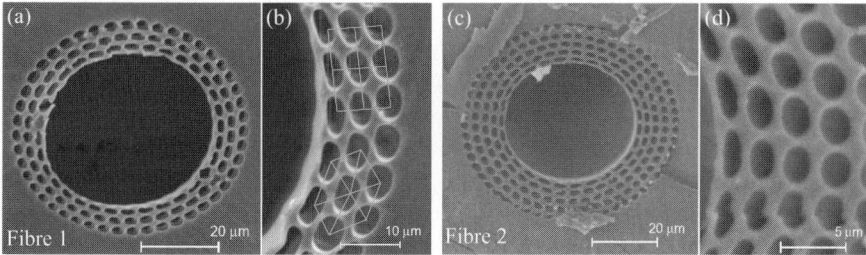

Fig. 8.3. Examples of the structure of bandgap fibres produced using drilled preforms. (a), (b) labelled as Fibre 1 and (c), (d) labelled as Fibre 2 were drawn from the cane shown in Fig. 6.10. The change of the local lattice of holes from square to hexagonal and the resulting diamond- and Y-shaped interstitial regions are indicated in (b) and the chirped thickness of the solid rings is observed in (d).

entire structure (core and cladding holes) was pressurised to increase the air-fraction in the cladding and reduce the thickness of the high-index features in the cladding without requiring the fibre to be drawn to a smaller diameter (see also Fig. 6.10 of Chapter 6). This resulted in the fibre with a "kagome" lattice (see [Benabid et al. 2002, Couny et al. 2006]) shown in Fig. 8.4(b). Alternatively, it was possible to avoid the use of pressurisation by using capillaries with thinner walls. A close up SEM image of the resulting lattice structure is shown in Fig. 8.4(c) [Argyros, unpublished]. The features in the cladding of a kagome lattice originate from both the stacked hollow tubes themselves as well as from the air voids between them (the interstitial holes), resulting in a typical "Star of David" pattern and an air fraction greater than 90%.

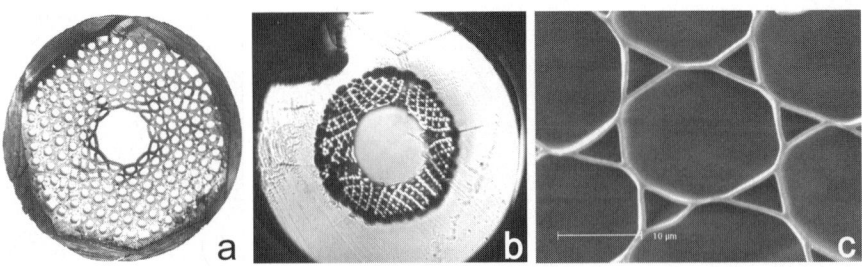

Fig. 8.4. (a) Cane of 12 mm diameter drawn from stacked preforms similar to the one shown in Fig. 5.4, (b) the resulting fibre as drawn from the cane shown in (a), and (c) a SEM close-up of a kagome-like lattice structure in a fibre drawn from a preform using thinner capillaries [Argyros, unpublished]. This fibre has an outer diameter of 470 μm, a core diameter of 35 μm, a pitch of 19.2 μm and a bridge thickness of 570 nm.

8.3 Transmission Properties

The various bandgap fibres were characterised by launching light from a supercontinuum source [Wadsworth et al. 2004] into the hollow core using a microscope objective lens. The output ends of the fibres were imaged using a second objective onto a CCD camera and guidance in the hollow core was observed. Note that guidance in a hollow core alone can be observed in simple hollow waveguides but this produces a characteristic blue colouration [Issa et al. 2003] which was not observed here (see Section 2.5). Photonic bandgap guidance is characterised by transmission only in discrete wavelength bands, which correspond to the bandgaps (see Section 3.1). The spectral output of the fibres was examined by imaging the end face onto a spectrum analyser through a spatial filter which selected only the light form the core. An additional characteristic of photonic bandgaps is that as the size of the microstructure is reduced, the bandgaps – and the resulting transmission bands – shift to shorter wavelengths.

Several samples of Fibre 1 as shown in Fig. 8.3(a) were drawn to different diameters and tested. The transmission spectra are shown in Fig. 8.5. For each sample, transmission in discrete wavelength bands was observed, as expected. Furthermore, these wavelength bands scaled with the fibre diameter (and the microstructure size). The shift in the transmission bands to shorter wavelengths as the structure size decreased was evident in the change of colouration of the output light.

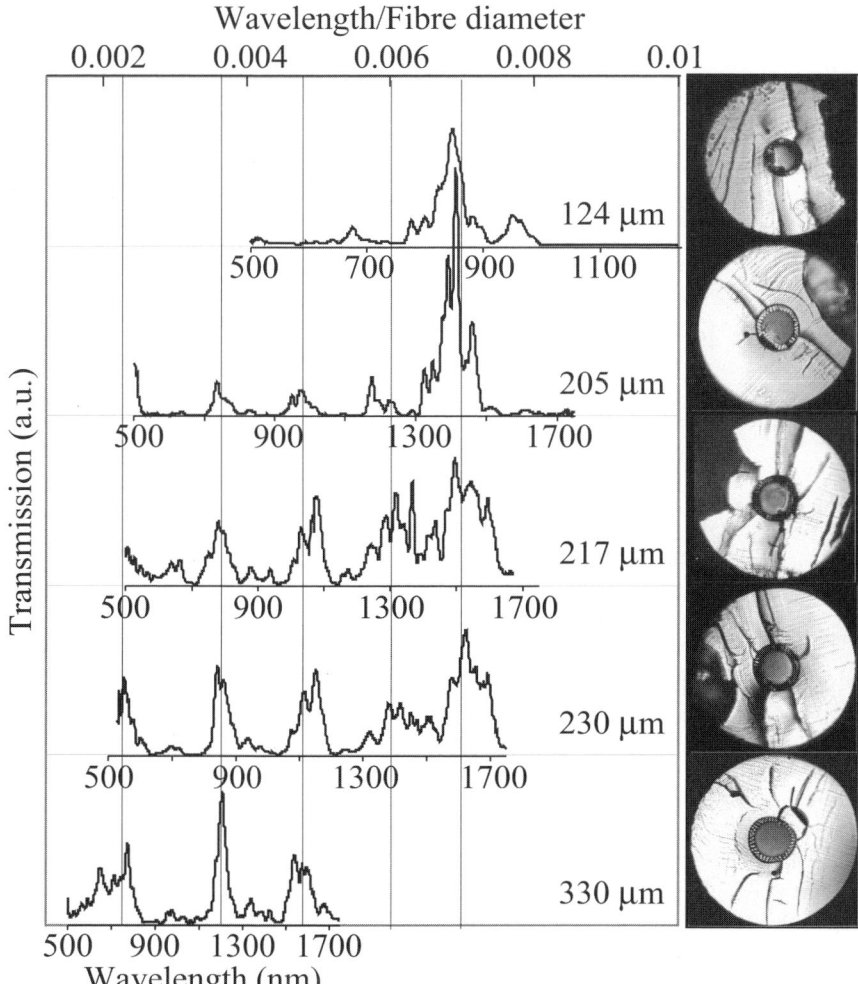

Fig. 8.5. Transmission (linear scale) through samples of Fibre 1 [Fig. 8.3(a)] drawn to different diameters (length ≈ 30 cm) shown as a function of wavelength and scaled wavelength. The insets show the different colours guided in the hollow core, changing through red, green, blue, blue-violet and yellow (= red + green) as the structure size decreases and the bandgaps shift to shorter wavelengths. After Argyros et al. [2006c]. See also colour plate at front of book.

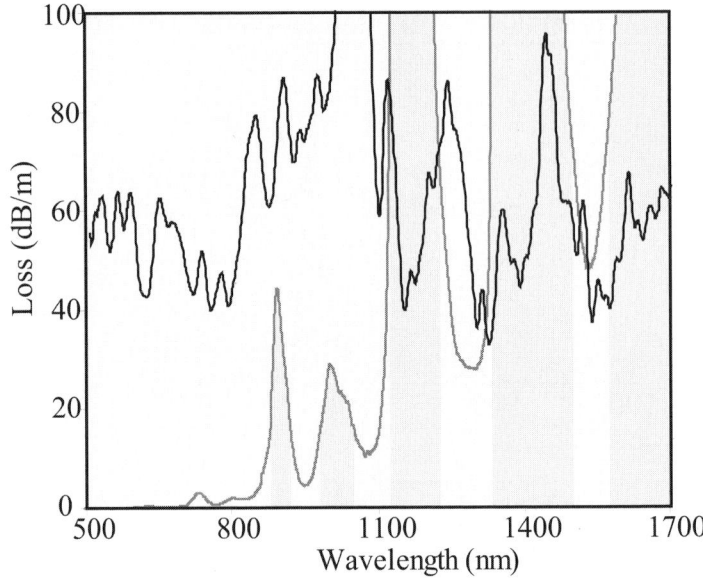

Fig. 8.6. The black curve shows the loss of a sample of Fibre 1 from Fig. 8.3(a) and the grey curve shows the material absorption of PMMA with absorption peaks highlighted. After Argyros et al. [2006c].

A significant advantage of HC-mPOF over conventional POF is its ability to transmit light with a loss below the material absorption. Indeed, with appropriate design, the transmission loss can be virtually independent of the material.

The transmission loss of the ring-structured Fibre 1 of Fig. 8.3(a) was measured by the cut-back method. The loss spectrum is shown in Fig. 8.6 superimposed on the PMMA material absorption. The lowest loss observed was 31 dB/m at $\lambda = 1300$ nm. This value is relatively high due to the heterogeneous cladding and small number of rings, both of which increase the confinement loss. Significantly however, the transmission loss remains below the material loss in the three transmission bands in the near infrared, such as at wavelengths above 1600 nm where the material absorption reaches values of 3000 dB/m. In this wavelength region, the transmission loss of ~ 60 dB/m, indicates an effective loss reduction of 50×, corresponding to less than 2% of the light being guided in the polymer. Numerical modelling of the ideal structure fibres showed that the effect of the material loss was reduced by a factor of 100 (compared to a solid-core fibre) for a fibre with a 5 μm core diameter and by a factor of 1000 for a 10 μm core diameter [Argyros et al.

2006a]. However, such scaling up of the core increases the number of surface modes, as will be discussed in the next Section.

The mPOF with the kagome lattice shown in Fig. 8.4(c) had two wide transmission bands spanning 480 to 650 nm and 950 to 1700 nm which shifted proportionally with the fibre diameter as expected [Argyros et al. 2006a]. The lowest loss was 13 dB/m measured at 580 nm and 1360 nm (for two different fibre diameters). In the infrared (specifically 1120 – 1680 nm), the transmission loss was up to 100× smaller than the material loss, the remaining loss being attributed to structural imperfections and confinement loss.

8.4 Issues Affecting The Performance Of Bandgap Fibres

8.4.1 The relative importance of the structural dimensions

Photonic bandgap fibres have generated enormous interest since they were first proposed, but have proved difficult to make, with very few groups in the world successfully achieving guidance in air. Understanding the physical processes underlying photonic bandgap guidance has been essential in eluci-dating why their fabrication is so difficult (see Section 3.1). The concentric cylinders for Bragg fibres, and the "rods" for the two-dimensional structures are high index regions that are considered as small resonators [Abeeluck et al. 2002, Litchinitser et al. 2002], and it is the regularity of these regions that is most significant in producing bandgap guidance. By contrast, the exact spac-ing of the high index inclusions is much less important, although it becomes more significant for longer wavelengths [Abeeluck et al. 2002]. Experiments with low-index bandgap fibres [Argyros et al. 2005a,b] have clearly demon-strated that bandgap guidance can be achieved in fibres where the structure is quite deformed (see Section 3.1).

More important than the exact periodicity of the array of rods are the size of the high index inclusions and the widths of the bridges connecting them. Variations in the size of the high index inclusions will broaden the range of allowed modes in the cladding: in particular, broadening the range of effective indices that can propagate into the cladding. Similarly, the width of the bridges determines the magnitude of the coupling between the rod modes. This also broadens the cladding bands. When this happens, the intervening bandgaps shrink; if the broadening is too great they may even disappear.

From a fabrication perspective, the need for high index rods with very thin connecting bridges turns out to be the major challenge, particularly for silica fibres where surface tension tends to promote hole collapse during the draw (see Chapter 6).

8.4.2 Role Of Surface Modes

The other major challenge in making hollow-core bandgap fibres is to control the interaction with surface modes such as shown in Fig. 8.7. Surface modes are known to be associated with the termination of periodic structures, and occur even when the structure is "perfect". Perturbations to a perfect structure at the core/cladding interface may increase the number of surface modes. Such modes are problematic in bandgap fibres because their location means that they overlap spatially with the core mode. If the effective index of the core mode is close to that of the surface mode, light can be coupled out of the core (see Fig. 8.8). Surface modes can also indirectly couple light to highly lossy radiation modes, which would normally not interact with the core due to their spatial separation [West et al. 2004].

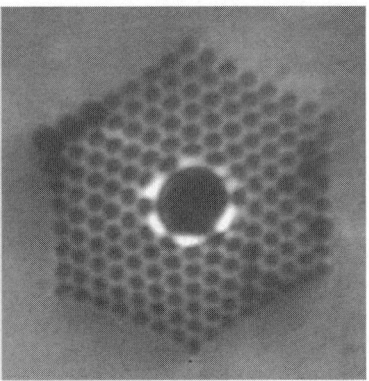

Fig. 8.7. The solid region surrounding the core of a hollow core fibre can support modes, as shown here, which can couple light out of the core. The impact of such surface modes can be reduced with appropriate design of the region around the core. See also colour plate at front of book.

While it is impossible to remove surface modes, it is possible to modify the design of the core/cladding interface so that they lie outside the bandgap. Designs that have been proposed for this purpose terminate the cladding array structure "correctly". In the most commonly used hexagonal array structure, the correct termination preserves the bridges joining the rods as interpenetrating "fingers" (see Fig. 8.9) at the interface [West et al. 2004]. These designs however are extremely difficult to make by capillary stacking. The large range of fabrication methods available for mPOF preforms however, offers much greater scope for fabricating these structures.

8.4.3 Ultimate Loss Limits For Hollow-Core MPOF

The performance HC-mPOF fabricated to date has been limited by deformations to the structure and confinement loss. It is however instructive to consider theoretically what the limitations to their performance would be, given improved fabrication.

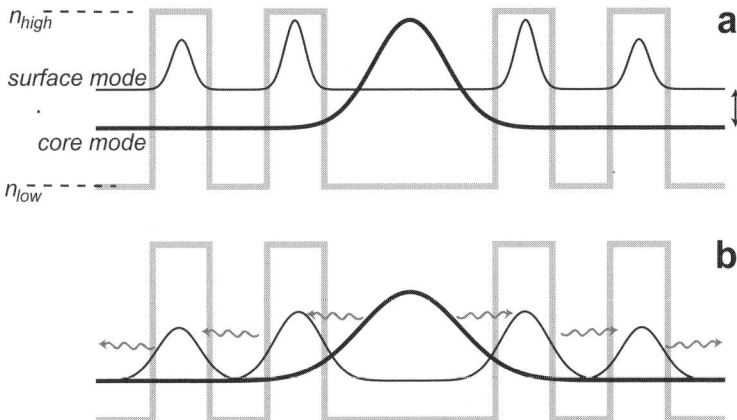

Fig. 8.8. The high index regions around a low index core may support modes that have a strong impact on fibre transmission. If the surface and core modes have very different effective indices as in (a) they will not interact. If they have a similar effective indices and a degree of spatial overlap, as in (b), they will cause coupling from the core to the cladding.

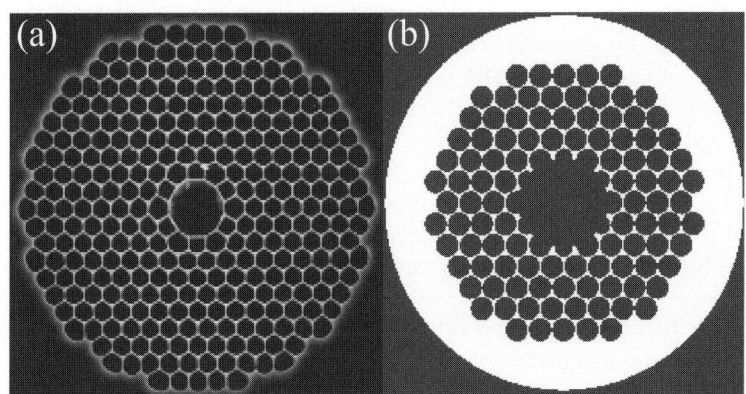

Fig. 8.9. (a) Example of a hollow-core fibre made using capillary stacking. The core-cladding interface exhibits a discontinuous termination of the periodic structure. (b) Schematic of a hollow-core fibre with a 'correct' termination showing interpenetrating fingers between the holes. *Image courtesy of Crystal Fibre A/S, Blokken 84, 3460 Birkerod, Denmark http://www.crystal-fibre.com*

In silica bandgap fibres, it has been shown that the ultimate loss limitation is due to scattering from roughness in the form of surface capillary waves. These are undulations frozen into the glass during the draw process [Roberts et al. 2005]. The impact of surface roughness can be reduced by increasing the core size. This reduces the relative intensity of the mode at the core/cladding

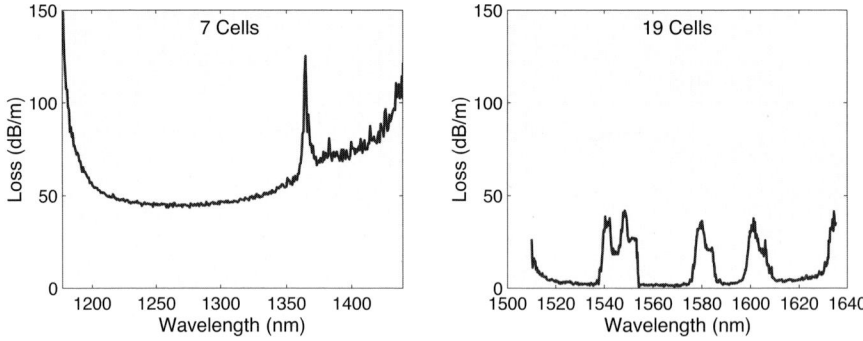

Fig. 8.10. Increasing the size of the core in a bandgap fibre reduces the overall loss by decreasing the intensity at the core/cladding interface, but increases the number of surface modes which fragment the bandgap. The core of (a) is formed by removing 7 capillaries, while 19 capillaries have been removed in fibre (b). After Russell [2006] (©[2006] IEEE).

interface, and hence the scattering from surface capillary waves. While this is effective in reducing the overall loss, it is not a complete solution, as the increase in the circumference of the core also increases the density of surface modes. The presence of these additional modes may fragment the bandgap. The choice of core size therefore involves trade-offs that depend on the specifications of the optical system [Humbert et al. 2004]. A comparison of experimental values of the loss for fibres with a 7-cell core and a 19-cell core are shown in Fig. 8.10.

In silica bandgap fibres, the impact of surface capillaries waves has been used to determine the ultimate transmission limits of these fibres [Roberts et al. 2005]. The same analysis can be applied to PMMA based bandgap fibres.

Taking all other parameters to be equal, the effects of surface capillary waves can be determined by the ratio of the processing temperature T_g to the surface tension γ of the material. The values are approximately $T_g = 1900$ K and $\gamma = 0.3$ N/m for silica, and $T_g = 400$ K and $\gamma = 0.032$ N/m for PMMA. Hence the effects would be expected to be twice as large for a PMMA fibre compared to an identical silica HC-MOF, due to the lower surface tension for the polymer. The predicted loss is for a bandgap fibre with a hexagonal arrangement of holes in the cladding, where the core is formed by the omission of a number of unit cells (holes) [Roberts et al. 2005]. The losses for fibres with a 7-cell and a 19-cell core are shown in Fig. 8.11. The polymer material loss cannot be neglected and the assumption made here for its inclusion is that 1% of the light will be present in the polymer for the smaller (7-cell) core and 0.1% for the larger (19-cell) core, based on the relevant experimental and theoretical predictions Humbert et al. [2004].

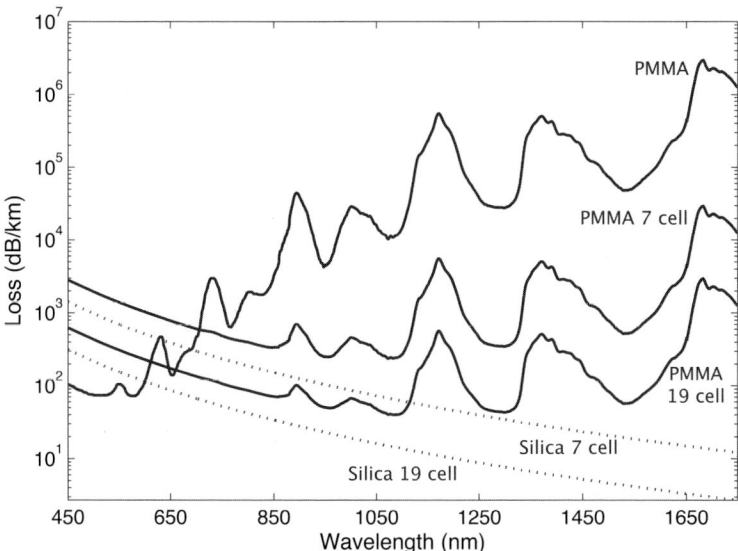

Fig. 8.11. The material loss of PMMA, the predicted loss of a HC-mPOF with a hexagonal structure and the loss of a silica HC-MOF (curves calculated by Argyros et al. [2006b] from the values presented by Roberts et al. [2005]).

Surface capillary waves dominate the loss for short wavelengths while the material absorption dominates at longer wavelengths, resulting in a qualitatively different behaviour to the silica fibres. Significant improvements are achieved compared to the material absorption, and the overall minimum loss is reduced to 50 dB/km at 1100 nm, compared to the current value of 150 dB/km at 650 nm. However, the scattering loss increases for the visible to above that of the bulk material. New transmission windows for PMMA fibres may thus be enabled at 700 to 1130 nm, 1230 to 1330 nm and 1470 to 1600 nm, potentially enabling the use of telecommunications related infrastructure and devices. To operate at these wavelengths, the contribution of the material loss would need to be minimised as much as possible and the fibre core would need to be large. This is compatible with current POF standards which use large-core multimode fibres, and is in fact desirable given the expected $1/r^3$ scaling of the loss with core radius [Johnson et al. 2001]. Reductions to the material loss by as much as a factor of 10^5 have been predicted for "Omniguide" Bragg fibres with large (100 μm) cores [Johnson et al. 2001].

Hollow-core mPOF extends POF technology to a far wider choice of polymers. Transparency is not as strict a requirement as for solid-core fibres, allowing materials that are thermally, mechanically or chemically optimal to be used. For example, one problem with the deployment of POF for the automotive industry is their low temperature stability. The automobile industry requires that polymer fibres made are able to withstand sustained exposure to temperatures above that achievable with PMMA. At present, mPOFs made from PMMA are limited to operating below about 120 °C while many automotive applications require that mechanical integrity and optical performance be maintained until at least 150 °C. Hollow-core mPOF could be produced using a more thermally stable polymer such as amorphous Teflon, while the additional requirements of flexibility and large fibre diameter preclude the use of silica.

8.5 Applications Of Photonic Bandgap Fibres

The impact and application of bandgap fibres will be determined by their loss performance. By guiding light in air, these fibres offer significant advantages. Air transmission of light is free of the limitations that occur in all other optical systems – to the extent that air approximates a vacuum it will have no absorb, dispersion or non-linearity. In practice, the presence of water vapour limits the performance.

The decrease in transmission loss for silica bandgap fibres that has occurred since their inception is impressive, though the ultimate loss limit (Section 8.4.3 is still considerably above those of conventional optical fibres. For hollow-core mPOF the situation is quite different, and the ultimate limit is expected to be *below* the loss of conventional polymer fibres (see Fig. 8.11).

There are a number of wavelength ranges (such as the mid-infrared and UV) for which it is difficult to produce low loss solid optical fibres from any material. Thus hollow-core guidance offers a potential advantage for applications in these wavelength range. Similarly, high-intensity transmission is required in a number of industrial and medical applications. Even when the absorption is very small, this can still cause problems if the intensity is sufficiently high, so guidance in air would be clearly preferable.

With HC-mPOF it has become possible to guide light in any transparent low-refractive index material, such as a gas or a liquid. Some of the most exciting work being carried out in hollow-core fibres recently has been in the area of gas based non-linear optics [Benabid et al. 2002, 2005], where they offer an unusual combination of long interaction lengths, good beam quality and high-intensity transmission at low power [Russell 2003]. From a sensing perspective, most samples of biological interest are aqueous, and maintaining a strong interaction between the solution and light has always been difficult. Again, bandgap fibres offer a cheap and practical solution [Fini 2004, Cox et al. 2006].

8.6 Challenges And Future Directions

The development of hollow core mPOF is potentially one of the most significant that the technology has to offer. The absorption loss of polymers has long limited their use in fibres, even when their mechanical properties give them a substantial advantage. Hollow core fibres circumvent this problem to a large degree, by allowing most of the intensity to travel in air rather than the polymer. In principle this enables the use of less transparent polymers (which may have for example better thermal stability) and the guidance of "difficult" wavelengths, those that fall within the absorption region of the material. The mid-infrared is one such region, with important medical applications.

In practice of course, reaping these benefits of hollow core fibres requires improvements in the fabrication. The measured losses are still well beyond the expected best performance. Achieving better losses will require both uniformity of the high index regions, thinner bridges to prevent coupling, and possibly lower surface roughness. A particularly interesting challenge will be to make larger cores. Most traditional applications of POF require large cores, and the effect surface roughness is diminished in this case. The trade-off in the hexagonal array designs, more surface modes, is potentially controllable by using better core/cladding interface structures. The best such structures are difficult to make by capillary stacking, but could be fabricated using techniques such as extrusion or moulding. An alternate approach of course is to use Bragg fibres.

Finally, hollow core fibres are notable for allowing transmission not just in air, but in other low index materials. In silica hollow core fibres this has lead to some truly impressive progress in nonlinear effects in gases. A less explored area is aqueous sensing, which for has numerous important applications in the medical and environmental fields. Using a waveguide geometry has allows much longer path lengths to be used, and far better collection efficiency. There is enormous scope for progress in this area.

References

Abeeluck, A K, Litchinitser, A N, Headley, C, and Eggleton, B (2002). Analysis of spectral characteristics of photonic bandgap waveguides. *Optics Express*, 10(23):1320–33.

Argyros, A, Birks, T A, Leon-Saval, S G, Cordeiro, C M B, Luan, F, and Russell, P St J (2005a). Photonic bandgap with an index step of one percent. *Optics Express*, 13(1):309–14.

Argyros, A, Birks, T A, Leon-Saval, S G, Cordeiro, C M B, and Russell, P St J (2005b). Guidance properties of low-contrast photonic bandgap fibres. *Optics Express*, 13(7):2503–11.

Argyros, A, Large, M C J, and van Eijkelenborg, M A (2006a). Progress and potential of hollow-core microstructured optical fibres. In *Proceedings of the International Plastic Optical Fibres conference*, Seoul, Korea.

Argyros, A, Manos, S, van Eijkelenborg, M A, Large, M C J, and Poladian, L (2006b). Applications of microstructured polymer optical fibres: hollow-core and graded-index fibres. In *Proceedings of the OptoElectronics and Communication Conference*, volume 11, pages 5D2–2, Kaohsiung, Taiwan.

Argyros, A, van Eijkelenborg, M A, Large, M C J, and Bassett, I M (2006c). Hollow-core microstructured polymer optical fibers. *Optics Letters*, 31(2):172–4.

Benabid, F, Couny, F, Knight, J C, Birks, T A, and Russell, P St J (2005). Compact, stable and efficient all-fibre gas cells using hollow-core photonic crystal fibres. *Nature*, 434(7032):488–91.

Benabid, F, Knight, J C, Antonopoulos, G, and Russell, P St J (2002). Stimulated Raman scattering in hydrogen-filled hollow-core photonic crystal fiber. *Science*, 298(5592):399–402.

Couny, F, Benabid, F, and Light, P S (2006). Large pitch kagome-structured hollow-core PCF. In *Proceedings of the European Conference on Optical Communications*, page Th4.2.4.

Cox, F, Argyros, A, and Large, M C J (2006). Liquid-filled hollow-core microstructured polymer optical fiber. *Optics Express*, 14(9):4135–40.

Fini, J M (2004). Microstructure fibres for optical sensing in gases and liquids. *Measurement Science and Technology*, 5:1120–8.

Humbert, G, Knight, J C, Bouwmans, G, Russell, P St J, Williams, D P, Roberts, P J, and Mangan, B J (2004). Hollow-core photonic crystal fibers for beam delivery. *Optics Express*, 12(8):1477–84.

Issa, N A, Argyros, A, van Eijkelenborg, M A, and Zagari, J (2003). Identifying hollow waveguide guidance in air-cored microstructured optical fibres. *Optics Express*, 11(9):996–1001.

Johnson, S G, Ibanescu, M, Skorobogatiy, M, Weisberg, O, Engeness, T D, Soljačić, M, Jacobs, S A, Joannopoulos, J D, and Fink, Y (2001). Low loss asymptotically single-mode propagation in large-core OmniGuide fibers. *Optics Express*, 9(13):748–79.

Koike, Y (1996). Status of POF in Japan. In *Proceedings of the International Plastic Optical Fibres conference*, volume 5, pages 1–8, Paris, France.

Litchinitser, N M, Abeeluck, A K, Headley, C, and Eggleton, B J (2002). Antiresonant reflecting photonic crystal optical waveguides. *Optics Letters*, 27(18):1592–4.

Roberts, P J, Couny, F, Sabert, H, Mangan, B J, Williams, D P, Farr, L, Mason, M W, Tomlinson, A, Birks, T A, Knight, J C, and Russell, P St J (2005). Ultimate low loss of hollow-core photonic crystal fibres. *Optics Express*, 13(1):236–44.

Russell, P St J (2003). Photonic crystal fibers. *Science*, 299:358–62.

Russell, P St-J (2006). Photonic-crystal fibers. *JOURNAL OF LIGHTWAVE TECHNOLOGY*, 24(12).

Shephard, J D, MacPherson, W N, Maier, P R J, Jones, J D C, Hand, D P, Mohebbi, M, George, A K, Roberts, P J, and Knight, J C (2005). Single-mode mid-IR guidance in a hollow-core photonic crystal fiber. *Optics Express*, 13(18):7139–44.

Temelkuran, B, Hart, S D, Benoit, G, Joannopoulos, J D, and Fink, Y (2002). Wavelength-scalable hollow optical fibres with large photonic bandgaps for CO_2 laser transmission. *Nature*, 420:650–3.

Wadsworth, W J, Joly, N, Knight, J C, Birks, T A, Biancalana, F, and Russell, P St J (2004). Supercontinuum and four-wave mixing with Q-switched pulses in endlessly single-mode photonic crystal fibers. *Optics Express*, 12(2):299–309.

West, J A, Smith, C M, Borrelli, N F, Allan, D C, and Koch, K W (2004). Surface modes in air-core photonic band-gap fibers. *Optics Express*, 12(8): 1485–96.

9

Graded-Index Microstructured Polymer Optical Fibres

> Zed: *"Bring out the Gimp".*
> Maynard: *"But the Gimp's sleeping."*
> Zed: *"Well, I guess you're gonna have to go wake him up now, won't you?"*
>
> Pulp Fiction (Quentin Tarantino, 1994)

In this chapter we describe the factors that limit the bandwidth of large-core multimode polymer fibres, including intramodal and intermodal dispersion. The latter is usually addressed by using a graded-index (GI) to equalise the group velocities of the fast and slow modes. While most GI fibres use differences in chemical composition to produce the index gradient, it is also possible to use a graded microstructure. Such fibres (denoted GImPOF) differ from conventional GI POF not just in having a microstructure, but also in having a much larger index contrast. We model the behaviour of a non-microstructured (i.e. azimuthally averaged) fibre with the same contrast. We present the modelling and experimental results for a GImPOF and assess its potential.

9.1 Material Dispersion

As has been discussed previously (see Section 2.3.1) there are a number of contributions to the dispersion properties of a fibre, including (intramodal) material and waveguide dispersion, and intermodal dispersion. Material dispersion is known to be relatively high for non-fluorinated polymers (see Fig. 3.10), particularly in their transparency region. The bandwidth performance of PMMA-based POF (see, for example Table 1.2) seems to reflect this, suggesting that material dispersion plays a major role in determining its bandwidth.

Microstructured fibres potentially offer considerable scope to compensate for material dispersion, because changes to the size and geometry of the microstructure can dramatically change the waveguide contribution to the total chromatic dispersion. In order to understand the significance of the microstructure, we need to determine whether its dominant role is in controlling the chromatic or the intermodal dispersion.

This question can be answered by considering the degree to which chromatic dispersion limits bandwidth. From Section 2.3 the optical bandwidth can be written:

$$f = \frac{1}{2\delta\tau} \tag{9.1}$$

where $\delta\tau$ is the maximum allowed pulse spread. Thus, if the bandwidth is 1 GHz then the maximum pulse spread is 0.5 ns.

On the other hand, if material dispersion D_λ is the only source of pulse spreading then

$$f = \frac{1}{2 D_\lambda L \Delta\lambda} \tag{9.2}$$

where $\Delta\lambda$ is the linewidth of the source and L is the length of fibre. Clearly then, the significance of material dispersion depends strongly on the linewidth of the excitation source. The general relationship between the linewidth of the source and the maximum bandwidth permitted by material dispersion is shown in Fig. 9.1.

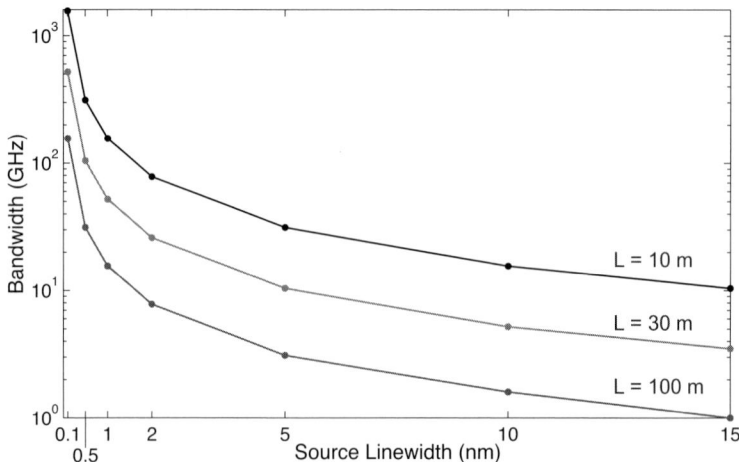

Fig. 9.1. Maximum bandwidth permitted by sources of different linewidth and fibres of length L assuming only material dispersion.

A variety of sources are used for POF including light emitting diodes (LEDs), laser diodes (LDs) and vertical cavity surface emitting laser diodes (VCSELs). LEDs are however unsuitable for high bandwidth applications because of their slow modulation rates. Laser diodes and VCSELs typically have linewidths in the range 1 to 4 nm. A suitable source for high-speed transmission in the visible is a Firecomms FC665V-002 VCSEL at 665 nm with a spectral width below 1 nm and a bandwidth exceeding 3 GHz. Using such a source, Eq. 9.2 and Fig. 9.1 reveal that material dispersion would limit the bandwidth over 100 m to 15.6 GHz, far in excess of the values experimentally obtained in conventional GI POF. Thus, in practice, material dispersion is not the limiting factor for GImPOF.

9.2 Intermodal Dispersion

The usual method to minimise intermodal dispersion is to use a graded-index refractive index profile. This structure allows the group velocities of the fast and the slow modes to be approximately equalised. In the ray model, we can view this as allowing a lower refractive index to compensate for the longer optical path length of higher order modes.

A power-law graded-index profile is given by

$$n^2(r) = n_{co}^2 - [n_{co}^2 - n_{min}^2]\left(\frac{r}{a}\right)^g, \qquad 0 \le r \le a \qquad (9.3)$$

where a is the radius of the core. If there is no material dispersion the optimal index profile has $g = 2$ [Snyder and Love 1983, Olshansky and Keck 1976]. The inclusion of material dispersion slightly alters the optimal value of g [Ishigure et al. 1996, 2003].

Extensive literature is available on conventional graded-index fibres and their behaviour is, at least theoretically, well understood. The same cannot be said for microstructured GI fibres. As discussed in Section 3.3, single-mode MOFs differ significantly from conventional single-mode fibres, and a similar situation seems to hold for their multimode counterparts. This poses significant challenges. To understand single-mode behaviour one only has to evaluate one or two modes, while for GI MOFs one may need to consider several thousand, a task that if done rigorously requires the use of a super-computer.

In order to develop some intuition about GI MOFs, we chose to consider a circular structure that had similar properties. Apart from the presence of the microstructure, one feature of GI MOFs that is quite different from conventional GI fibres is that the index difference is far larger. In addition, the cladding index in GI MOFs is the same as that of the core, which leads to confinement loss even for a non-microstructured (azimuthally-averaged) graded-index region (see Fig. 9.2).

9.3 The Effect of the Graded-Index Microstructure

A microstructured GI fibre and its azimuthally-averaged equivalent (see Fig. 9.2) were compared. Both have a core radius of 135 μm, and an approximately parabolic index profile (corresponding to $g = 2$). The minimum azimuthally-averaged refractive index, n_{min}, is 1.1. These parameters were chosen for two reasons. A computational study of azimuthally-averaged structures showed that $g = 2$ gave the optimal bandwidth [Manos 2006]. The same study suggested that n_{min} should be close to the polymer refractive index. However, it is difficult to fabricate relatively shallow graded profiles using the available methods.

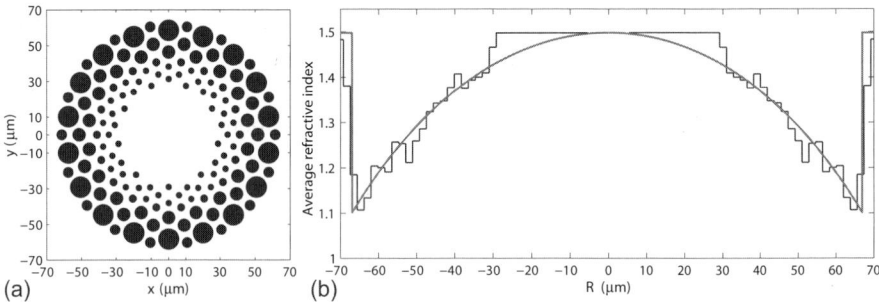

Fig. 9.2. Index profile for (a) the microstructured fibre and (b) its azimuthally-averaged counterpart used in the modelling. The parabolic approximation to the index profile is also shown in (b).

Conventional GI structures have generally been modelled by approximating the distribution of mode velocities [Olshansky and Keck 1976, Yabre 1976]. Microstructured GI fibres are such a new field that no useful approximation for the mode distribution is known, and numerically predicting the bandwidth performance requires the explicit evaluation of all propagating modes and their corresponding group velocities. This sort of extensive modelling has previously been done using scalar approximations for large-core traditional multimode fibres [Eguchi 2004, Eguchi and Horinouchi 2004, Eguchi 2005]. In our case, the high refractive index contrasts meant that the weak guidance approximation was invalid, and a vectorial simulation was required. Modelling of both structures was done using the Adjustable Boundary Condition method (see Section 4.2.2 and [Issa and Poladian 2003]) at $\lambda = 650$ nm, and taking the material to be PMMA.

Confinement loss

For conventional fibres, the number of bound modes and their corresponding cutoff is unambiguous. In microstructured fibres, cutoff is determined by the relationship between the confinement loss and details of the geometry (see Section 2.1.3 and 7.2.4).

For the two fibres compared here the effective cutoff was determined numerically: a very large number of modes were found and a sharp increase in confinement loss was used to locate the cutoff.

A comparison of how many modes are well confined, their range of group index and related properties of both fibres is given in Table 9.1. The confinement losses of the modes for the microstructured fibre are shown in Fig. 9.3 and they reveal a well defined cutoff: modes with $n_{\mathrm{eff}} < 1.484$ exhibit qualitatively higher confinement losses. Note that the cutoff index is very much higher than n_{min}.

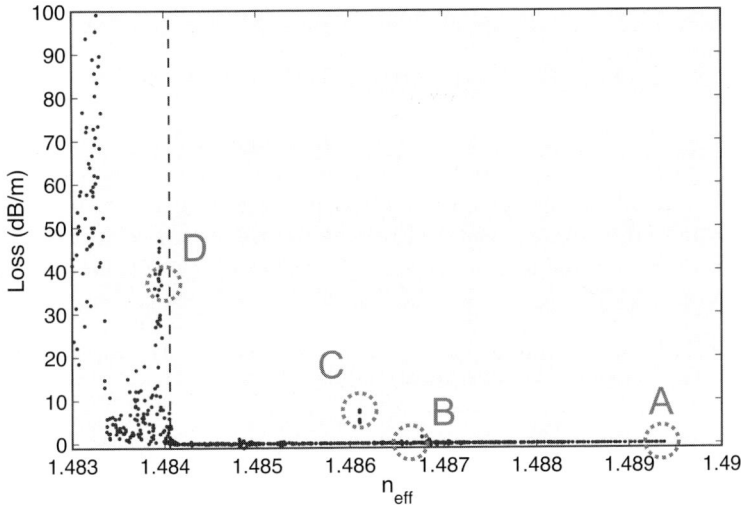

Fig. 9.3. Cutoff behaviour of the 135 micron core GImPOF shown in Fig. 9.2. The intensity profiles of typical modes from the groups identified as A-D are shown in Fig. 9.5

Intermodal dispersion

The relationship between n_{eff} and n_{g} for the azimuthally-averaged and microstructured GI fibres is shown in Fig. 9.4. The systematic variation of group index with phase index is used as the basis for estimating the intermodal dispersion. In the absence of power-mixing, the difference between the minimum and maximum group index would determine the degree of pulse spreading (as calculated in the last two rows of Table 9.1). Power-mixing reduces the pulse spread by effectively compressing the range of group velocities and also changing the dependence on fibre length (see Section 4.3).

Comparison of the microstructured fibre and its azimuthally-averaged equivalent clearly indicates that the microstructure is a critical determinant

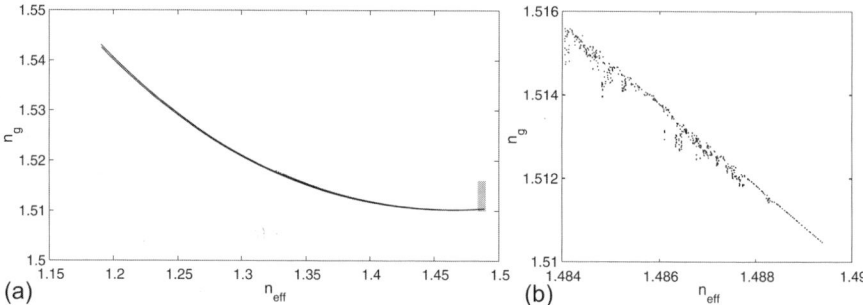

Fig. 9.4. The relationship between n_g and n_{eff} for the (a) azimuthally-averaged and (b) microstructured GI index profiles shown in Fig. 9.2. The gray box on (a) shows the region covered in (b).

of the optical behaviour. In other words, the properties of a microstructured GI fibre are not well predicted by assuming a homogenised structure. The results also highlight the following difference in the operation of GI fibres. Conventional GI fibres are designed to equalise the group velocities, whereas GImPOF appear to reduce the range and number of low-loss modes.

Table 9.1. Comparison of microstructured and non-microstructured GI fibres.

	Non-microstructured	Microstructured
Cutoff n_{eff}	1.19061	1.48405
δn_{eff}	0.2981	0.005328
N_{modes} above cutoff	2.4×10^4	1.4×10^3
Minimum n_g	1.51024	1.51051
Maximum n_g	1.54304	1.51511
δn_g	0.03281	0.0046
Intermodal dispersion (ps/m)	109.5	15.4

Anomalous modes

The intensity profiles of four selected modes indicated by letters in Fig. 9.3, are shown in Fig. 9.5. The mode indicated by C is anomalous in that its effective index is in the range of low loss but it has a confinement loss typical of modes below cutoff. The interlaced rings of small and large holes in a GImPOF are an attempt to approximate the smoothly varying parabolic profile in conventional GI POF. However, at certain radii, even after azimuthal-averaging, narrow annuli of slightly higher average index remain and anomalous modes such as C can be localised in such regions. While it is possible to smooth out these index fluctuations by improvements to the design, the bandwidth is

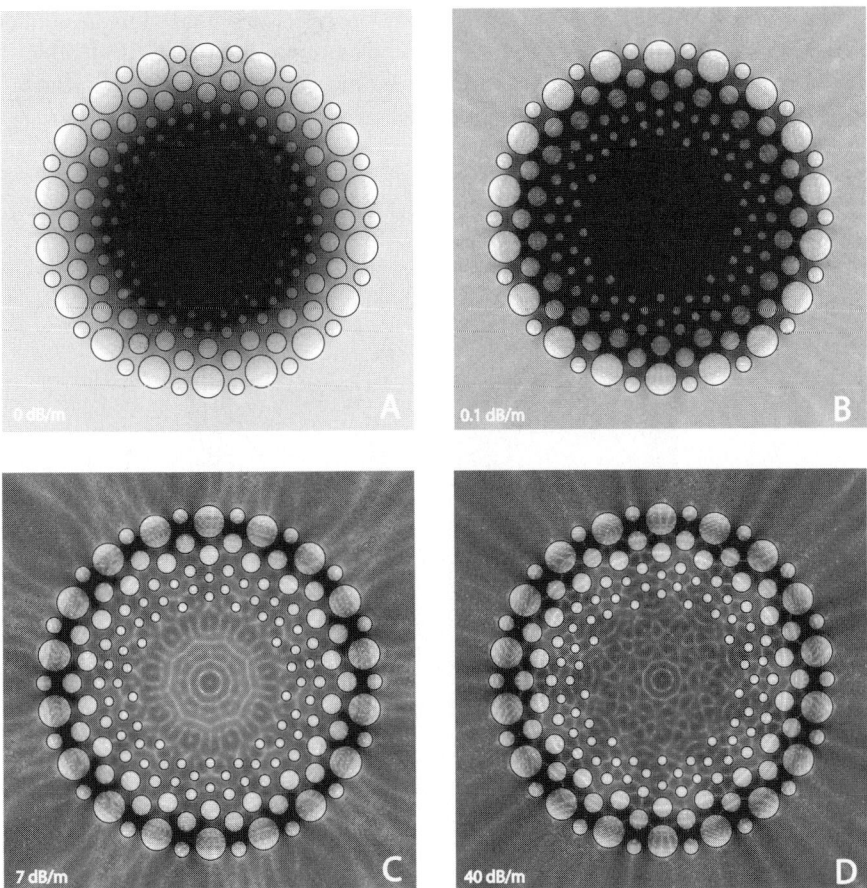

Fig. 9.5. Intensity profiles of the four modes identified in Fig. 9.3 are shown on a logarithmic gray-scale where black is the most intense. **A** is the fundamental mode. **B** corresponds to a slightly lossier mode which extends more into the hole structure. **C** is an anomalous mode which is localised to a narrow region within the hole structure. **D** is a mode just below the cutoff.

determined by the collective properties of a large number of modes and would not be significantly affected by a handful of anomalous modes.

Differential mode attenuation

Intermodal dispersion is not the only way that the fibre structure can impact on bandwidth. Another potential factor is differential mode attenuation (DMA). Large DMA can improve bandwidth, though potentially at the cost of increased overall loss. By removing the lossy high-order modes, the power distribution can be skewed towards lower order modes, and thus a smaller range of n_g. This effect has been seen in conventional multimode POF [Ishigure et al. 2000] but may be more pronounced in microstructured fibres [Barton et al. 2003] due to the role of confinement loss.

9.4 Experimental Characterisation Of GImPOFs

Over a five year period, several fibres were drawn from preforms having the design in Fig. 9.2. Four such fibres are shown in Fig. 9.6.

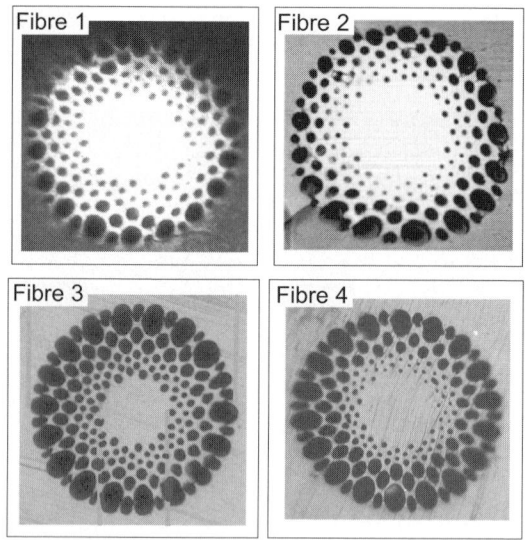

Fig. 9.6. Four different GImPOFs manufactured over a five year period based on the design in Fig. 9.2. Apart from variations in the hole shapes the main difference is the final size of the inner solid core. Fibre 2 after van Eijkelenborg et al. [2004].

The small variations in geometry (e.g. hole sizes, shapes and the size of the inner core) are due to the use of slightly different draw conditions. More important, however, are the changes to the surface quality as a result of improvements to the fabrication process. These fibres were made and measured over a long period of time and the available lengths of fibre suitable for measurement have been steadily improving. For example, the bandwidth measurement for the oldest fibre (Fibre 1) is only a bound since no dispersion was detectable over the short length of fibre available.

The measured loss and bandwidth of these four fibres are shown in Table 9.2. The actual length used during the measurement is given as well as an extrapolation of the bandwidth to a common length of 100 m. The extrapolation is given as a range: the smaller limit assumes no mode-mixing (or equivalently a long equilibrium length) the upper limit assumes equilibration under mode-mixing has occurred.

Table 9.2. Summary of the attenuation, bandwidth and length of fibre used during characterisation. *Fibre 1, measured at the Australian National University [Barry Luther-Davies, unpublished result]. Fibres 2 and 3 measurement at the Polymer Optical Fibre Applications Centre (POFAC), Nürnberg, Germany using setup shown in Fig. 7.15(a). Fibre 4 measured at the University of Auckland using setup shown in Fig. 7.15(b).*

Fibre	Attenuation (dB/m)	BW (GHz)	Fibre length (m)	Extrapolated BW for 100 m (GHz)
1	4.7	18.9	1.2	0.23 to 2.1
2	0.8	5	10	0.5 to 1.6
3	0.4	10	25	2.5 to 5.0
4	–	13	28	3.6 to 6.9

Conventional wisdom suggests that improvements in bandwidth are possible only at the expense of transmission, through mode-mixing and DMA. However, the historical improvement in bandwidth from Fibre 1 though to Fibre 3 has been accompanied by an improvement in loss.

Fibres 3 and 4 were manufactured in the same way but were measured under different conditions. This allows us to draw conclusions about the relative role of material dispersion. The source used to obtain the measurements for Fibre 3 has a linewidth of 3.5 nm and thus there is a substantial contribution from material dispersion. The bandwidth of Fibre 4 was measured using a source with a linewidth of 0.01 nm. The use of a very narrow linewidth source allows the contribution of material dispersion to be reduced to the level where it can be neglected. The dependence of the bandwidth on the intermodal dispersion is not very strong for this design.

Table 9.3 shows a comparison of the two most recent fibres (Fibres 3 and 4) with conventional high-speed transmission fibres. The bit rates at 100 m are extrapolated conservatively assuming that power-mixing does not occur.

9.5 Challenges And Future Directions

While the results quoted are promising, significant challenges remain. One of the most compelling of these is to reduce the loss to a similar level to conventional GI fibres. Much of the mPOF loss improvement achieved in the past has come from better fabrication (see Section 7.2.5), and from thicker sleeving of the fibres to reduce microbending. One aspect of fabrication that has remained largely unchanged over several years is the drilling of the preform. This is a likely source of surface roughness, and may limit loss improvements until a cleaner fabrication method is used.

Table 9.3. Overview of recent high speed transmission experiments [Pedrotti 2006] along with the results for Fibre 3. SI and GI refer to step-index and graded-index respectively, and PF refers to perfluorinated polymer.

Bit rate (Gb/s)	Length (m)	Fibre type	Core diameter (μm)	λ (nm)
1	100	PF-GI	120	850
2.2	11.9	PMMA-SI	1000	780
2.5	100	PMMA-GI	420	647
2.5	550	PF-GI	170	840
5	**100**	**mPOF-Fibre 3**	**135**	**652**
7.2	**100**	**mPOF-Fibre 4**	**135**	**647**
7	80	PF-GI	155	930
11	100	PF-GI	130	1300

A further challenge is to understand the physical basis of the bandwidth performance. The relative role and interactions of the mechanisms (such as equalising the group velocities, mode-mixing and DMA) have not been completely unravelled. The task of modelling the complete system remains an extremely difficult one. Until this happens it is hard to know how to design and optimise such fibres (see Chapter 4).

References

Barton, G, van Eijkelenborg, M A, Henry, G, Issa, N A, Klein, K-F, Large, M C J, Manos, S, Padden, W, Pok, W, and Poladian, L (2003). Characteristics of multimode microstructured POF performance. In *Proceedings of the International Plastic Optical Fibres conference*, volume 12, pages 81–4, Seattle, USA.

Eguchi, M (2004). All propagation modes of large-core multimode optical fibers with an arbitrary core profile. *Optics Letters*, 29(10):1051–3.

Eguchi, M (2005). Numerical analysis of modal dispersion in graded-index plastic optical fibres. *Applied Optics*, 44(26):5544–8.

Eguchi, M and Horinouchi, S (2004). Finite-element modal analysis of large-core multimode optical fibres. *Applied Optics*, 43(10):2163–7.

Ishigure, T, Kano, M, and Koike, Y (2000). Which is a more serious factor to the bandwidth of GI POF: Differential mode attenuation or mode coupling. *Journal of Lightwave Technolgy*, 18(7):959–65.

Ishigure, T, Makino, K, Tanaka, S, and Koike, Y (2003). High-bandwidth graded-index polymer optical fiber enabling power penalty-free gigabit data transmission. *Journal of Lightwave Technology*, 21(11):2923–30.

Ishigure, T, Nihei, E, and Koike, Y (1996). Optimum refractive-index profile of the graded index polymer optical fiber, toward gigabit data links. *Applied Optics*, 35(12):2048–53.

Issa, N A and Poladian, L (2003). Vector wave expansion method for leaky modes of microstructured optical fibres. *Journal of Lightware Technology*, 22(4):1005–12.

Manos, S (2006). *Designing Fibre Bragg Gratings and Microstructured Optical Fibres with Genetic Algorithms*. PhD dissertation, School of Physics, The University of Sydney, Sydney, Australia.

Olshansky, R and Keck, D B (1976). Pulse broadening in graded-index optical fibres. *Applied Optics*, 15(2):483–91.

Pedrotti, K D (2006). Multi-gigabit transmission on POF. *POF Newsletter*, 15(3).

Snyder, A W and Love, J D (1983). *Optical waveguide theory*. Chapman and Hall, New York.

van Eijkelenborg, M A, Argyros, A, Bachmann, A, Barton, G W, Large, M C J, Henry, G, Issa, N A, Klein, K F, Poisel, H, Pok, W, Poladian, L, Manos, S, and Zagari, J (2004). Bandwidth and loss measurements of graded-index microstructured polymer optical fibre. *Electronics Letters*, 40(10):592–3.

Yabre, G (1976). Comprehensive theory of dispersion in graded-index optical fibres. *Journal of Lightwave Technology*, 18(2):166–77.

10

Bragg and Long Period Gratings in mPOF

Inventions have long since reached their limit, and I see no hope for further development.

Julius Sextus Frontius, Roman engineer, 1st century AD.

This chapter describes both Fibre Bragg Gratings (FBG) and Long Period Gratings (LPG) in mPOF, and the required theoretical basis to understand their operation. Gratings are a highly developed research area, and no attempt has been made to review the whole field. Excellent reviews are available, including [Othonos and Kalli 1999, Kashyap 1999]. The two most relevant related areas however are described here: gratings in MOFs and POFs. This work is significant in identifying applications where the material or waveguide properties of mPOF may offer new functionality. The chapter also describes the experimental techniques used to make mPOF gratings, and their performance characteristics. Finally, conclusions are drawn about how mPOF gratings may be improved and application areas where they may prove useful.

10.1 Introduction

The first Fibre Bragg Gratings were written accidentally in a germania-doped silica fibre using a high power argon ion laser [Hill et al. 1978]. During a non-linear optics experiment, light was increasingly reflected out of the fibre. The forward propagating wave and the wave reflected at the fibre end-face had set up a standing wave, and this modulation in intensity in turn caused a modification in the material properties of the glass and hence a permanent grating, now known as a "Hill grating".

Following this serendipitous result, a huge effort was put into fibre gratings: improving their fabrication (they are now all externally inscribed), obtaining complex profiles, optimizing their performance and incorporating them into devices. Compact and integrated into the fibre, FBGs have replaced bulk optics in a wide range of applications where they are used as reflecting elements, filters, dispersion compensators and sensing elements [Othonos and Kalli 1999, Kashyap 1999]. FBGs have also been written in POF, though the difficulty in obtaining single-mode fibres, and the prevalence of phase masks at 1550 nm has meant that they have been written in the infrared region, where the polymer transparency is low.

A new type of fibre grating was developed about a decade ago [Vengsarkar et al. 1996]. Long period gratings (LPG) couple light from forward propagating core modes to forward propagating cladding modes. Currently less developed than FBGs, LPGs also have a diverse range of applications, including gain-flattening, band-rejection filters, variable attenuators [Eggleton et al. 2001], band-pass filters [Choi et al. 2002], mode conversion [Lee and Erdogan 2001], pressure, temperature, and strain sensors [Han et al. 2003, Liu et al. 1999], tunable filters, and dispersion compensators [Stegall and Erdogan 2000]. LPGs have a much higher sensitivity to external perturbations than FBGs, and are thus particularly suited to sensing applications.

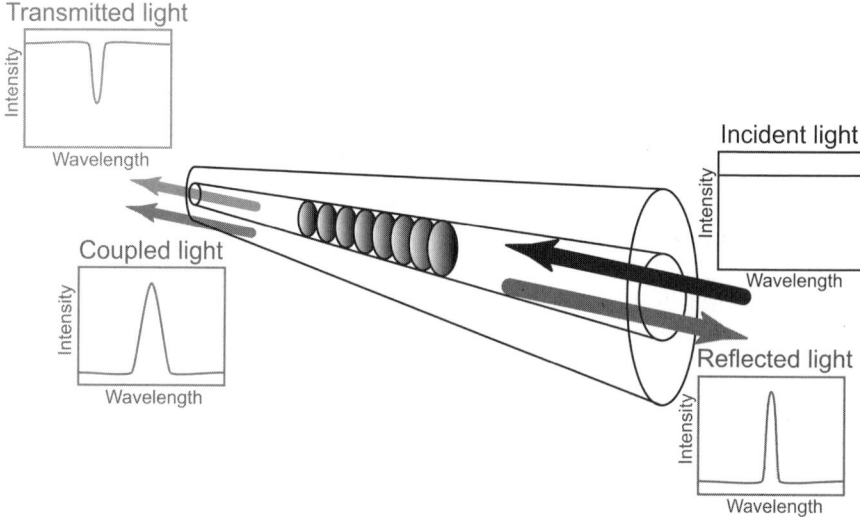

Fig. 10.1. A fibre grating is a periodic variation in the core of an optical fibre which causes coupling between modes. The incident and transmitted modes are shown with their correspondingly labelled spectra. A Bragg grating couples to a backward propagating mode as shown by the reflected spectrum, whereas a long-period grating couples to another forward propagating mode (often in the cladding) as shown by the spectrum labelled as coupled light. After Othonos and Kalli [1999].

Figure 10.1 shows a schematic of a fibre grating. In a Bragg grating, these are the forward and backward propagating core modes, causing light to be reflected, while in a long period grating the coupling is between forward propagating core and cladding modes. The properties of the grating depend on the periodicity, refractive index contrast and the envelope function. The effect of changing some of these features is illustrated in Fig. 10.2.

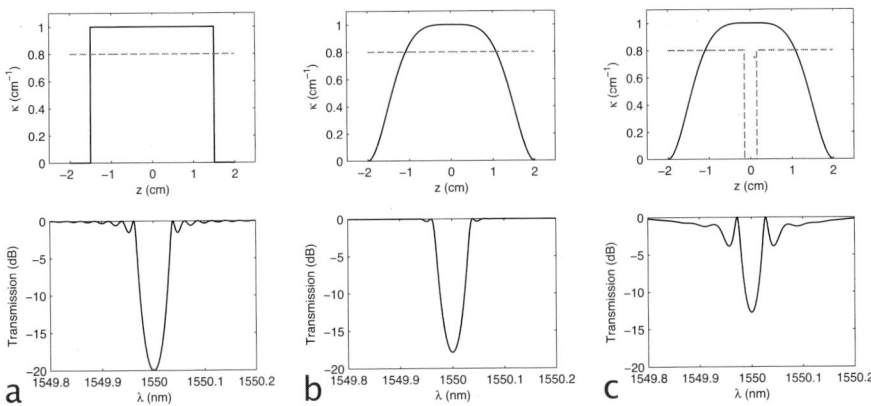

Fig. 10.2. The effect of changing some of the parameters of a FBG. The grating profiles (top) are shown with their corresponding transmission spectra (bottom). The grating strength is represented by the solid black line and the phase by the dashed red line. A uniform grating (a), results in a typical stop band spectrum. An apodised grating (b) suppresses the sidelobes of the spectrum. The introduction of two phase jumps (c), results in two transmission peaks within the stop band.

FBGs and LPGs are both deterministic perturbations to the fibre which cause coupling between the modes (see Section 4.3). In a FBG, the coupling is between a forward and a backward propagating mode (usually the fundamental core mode), while in a LPG the coupling is between two forward propagating modes (usually a core and a cladding mode). FBGs result in the reflectance of particular wavelengths from the grating, while LPGs cause selected wavelengths to be coupled from the core into the cladding, where they are ultimately lost through scattering and absorption.

The interaction of modes coupled by any grating must satisfy conservation of momentum:

$$\mathbf{k_1} + \mathbf{K} = \mathbf{k_2} \tag{10.1}$$

Here $\mathbf{K}$ is the wavevector of the grating with a direction normal to the grating plane and a magnitude of $2\pi/\Lambda$. Λ is the period of the grating. The modes coupled by the grating have wavevectors k_1 and k_2. In FBGs these wavevectors are equal in magnitude but opposite in sign, while in LPGs the magnitude differs but the sign is the same. The effective indices of the modes are given by n_{eff}^1 and n_{eff}^2 respectively, and the wavelength coupled by the grating is:

$$\lambda = \left(n_{\text{eff}}^1 \mp n_{\text{eff}}^2 \right) \Lambda \qquad (10.2)$$

By convention, n_{eff}^1 and n_{eff}^2 are positive and hence do not express the direction in which the mode is travelling. In a FBG, where the grating couples modes travelling in opposite directions, the effective indices add, while in an LPG they are subtracted. While the fundamental equations governing the spectral features of FBGs and LPGs are the same, there are differences associated with the strength of the grating. In LPGs all modes coupled by the grating are forward propagating and light will couple from one mode to the other as the length of the grating increases. Thus in a LPG, for a given fixed depth grating, there is an optimal grating length which will allow full transfer of energy from the core to the cladding mode [Daxhelet and Kulishov 2003]. At longer grating lengths the energy will begin to couple back into the core mode. In FBGs, by contrast, increasing the length of grating can only increase the reflectivity of the grating.

The effects of strain (ε) and temperature (T) have been widely studied in gratings. This is both because such perturbations may be problematic, changing the behaviour of a device, and because they allow strain and temperature to be accurately measured. The effect of either can be understood by differentiating Eq. 10.2, and rearranging [Allsop et al. 2002]. For simplicity, only the equation describing the effect of temperature is presented. The analogous equation however also describes the effect of strain, with the appropriate substitution of ε for T:

$$\frac{d\lambda}{dT} = \frac{\lambda}{\Delta n_{\text{gr}}} \frac{d(\Delta n_{\text{eff}})}{dT} + \frac{(\Delta n_{\text{eff}})^2}{\Delta n_{\text{gr}}} \frac{d\Lambda}{dT} \qquad (10.3)$$

where

$$\Delta n_{\text{eff}} = n_{\text{eff}}^1 \mp n_{\text{eff}}^2 \qquad (10.4)$$

and

$$\Delta n_{\text{gr}} = \Delta n_{\text{eff}} - \lambda \frac{d(\Delta n_{\text{eff}})}{d\lambda} \qquad (10.5)$$

The first term in Eq. 10.3 refers to temperature induced changes to the effective index of the modes, while the second term refers to the change in the periodicity of the grating. These changes can be related to the material and waveguide properties respectively.

Equation 10.3 also makes clear a fundamental difference between the operation of FBGs and LPGs. In FBGs there is effectively only one mode: the forward and backward propagating core mode, so Δn_{eff} becomes $2\, dn_{\text{eff}}^{\text{co}}/d\varepsilon$, where the "co" superscript denotes the core mode. In LPGs however, the core and cladding modes will be affected differently by the temperature. Equation 10.3 shows that the shift in wavelength is related to the differential change in the modes. This is significant because it means that with an appropriate choice of parameters, the first term in Eq. 10.3 can be made zero. Indeed, the sensitivity of LPGs can be widely varied by the use of appropriate design.

10.2 Fibre Bragg Gratings

10.2.1 FBGs In Microstructured Fibres

While there is an extensive literature on FBGs in conventional silica fibre, relatively little work has been reported on gratings in MOFs. This is probably due to the technical difficulties associated with the inscription, which generally involves focussing the writing beam into the core of the fibre. In MOFs, this is difficult because the holes around the core scatter the writing beam. Numerical simulations of plane-waves propagating transversally through the microstructure confirm the significance of scattering by the cladding, and show that the intensity reaching the core is not uniform [Beugin et al. 2006]. This has not prevented gratings being written, but higher intensity sources were needed to compensate for the loss of intensity due to scattering.

FBGs were first fabricated in a silica MOF with a small germanium-doped core [Eggleton et al. 1999] and later written in a pure silica microstructured fibre [Groothoff et al. 2003, Fu et al. 2005]. As would be expected, gratings in the pure silica case were considerably weaker, with strengths of about 14 dB and 10 dB respectively, compared to 40 dB in the doped fibre.

A number of applications have been explored for FBGs in silica-based MOFs, such as the reflective elements of fibre lasers [Søndergaard 2000, Caning et al. 2003], and in strain and temperature sensing [Frazão et al. 2005]. The novel properties in both these applications however were related more to the fibre than that of the grating. The dispersion properties of FBGs have also been used to enhance supercontinuum generation in MOFs [Li et al. 2005a], and more generally there seems considerable scope to use FBGs as an additional tool in the development of non-linear optical devices.

10.2.2 FBGs In Polymer Fibres

Unlike silica gratings, POF gratings have not been widely used, and the physical mechanisms behind grating formation remain unclear. A number of processes (photochemical, photomechanical or thermochemical) can result in the creation of POF gratings, depending on the power of the UV source and the inscription time.

The inscription process has two identifiable stages corresponding to low and high-index modulation gratings [Liu et al. 2003]. "Type I" (so named in analogy to silica) gratings are due to a UV-induced refractive index change in the fibre core due, for example, to cross-linking [Tomlinson et al. 1970] or photochemical transitions in dopants added to the core [Yu et al. 2004]. The reflection and transmission spectra are complementary, indicating that there is little or no additional loss, as might be associated with coupling to cladding modes or absorption. Type II gratings are thought to arise from damage at the core-cladding interface. They have broader spectral features than Type I gratings, and significant losses at short wavelengths. An unusual writing process has been identified when the gratings are written at low power: gratings appear but are erased by the UV beam, reappearing after the beam is switched off. The likely reason for this behaviour is competition between UV induced photochemical changes and thermal effects, which gradually disappear when the irradiation is turned off [Liu et al. 2004].

The motivation for making FBGs in POF has centred on the fact that the mechanical and thermal properties of polymer are very different to those of silica (see Section 1.4). Tuning of up to 73 nm has been reported for PMMA fibres [Xiong et al. 1999], whereas > 2 nm is typical for silica, although more complicated efforts combining compression and strain can achieve wider tuning of > 15 nm [Rocha et al. 2005]. This effect can be used either for strain monitoring, or to provide wavelength tuning for optical elements. An erbium-doped fibre laser was tuned over a 35 nm range using an external POF FBG [Liu et al. 2006].

The relatively high value of the thermal expansion coefficient and the thermo-optic coefficient result in much higher thermal sensitivity than is possible for silica. This fact has been exploited in temperature sensing with an 18 nm shift in the reflected wavelength being observed over a 50 °C range with no thermal hysteresis [Liu et al. 2001].

10.2.3 FBGs In Microstructured Polymer Fibres

Fibre Bragg Gratings were first written in mPOF using a UV inscription technique based on that used for silica fibres [Dobb et al. 2005, Webb et al. 2005]. A phase mask with a period of 1060.85 nm was used, designed to produce a FBG in silica at approximately 1536 nm. The source was a continuous wave helium-cadmium (He-Cd) laser with an output wavelength of 325 nm and a power of 30 mW. Two plano-convex cylindrical lenses with 10 cm focal lengths were used, one placed in front of the phase mask to focus the light down toward the core, and the other to expand the 1.8 mm diameter laser beam to cover approximately a 1 cm length of the mPOF. The experimental setup is shown in Fig. 10.3.

Although a standard grating writing technique was used, a number of modifications were required to produce satisfactory gratings. POF are less rigid than silica fibres and need to be supported from beneath to prevent the fibre shifting during the relatively long writing period. Successful inscription also required careful control of the separation between the phase mask and the fibre to ensure that the grating was written in the core and not in the microstructured cladding.

A factor which may be significant in writing FBGs in mPOF is the orientation of the microstructure with respect to the incident laser beam. Dobb

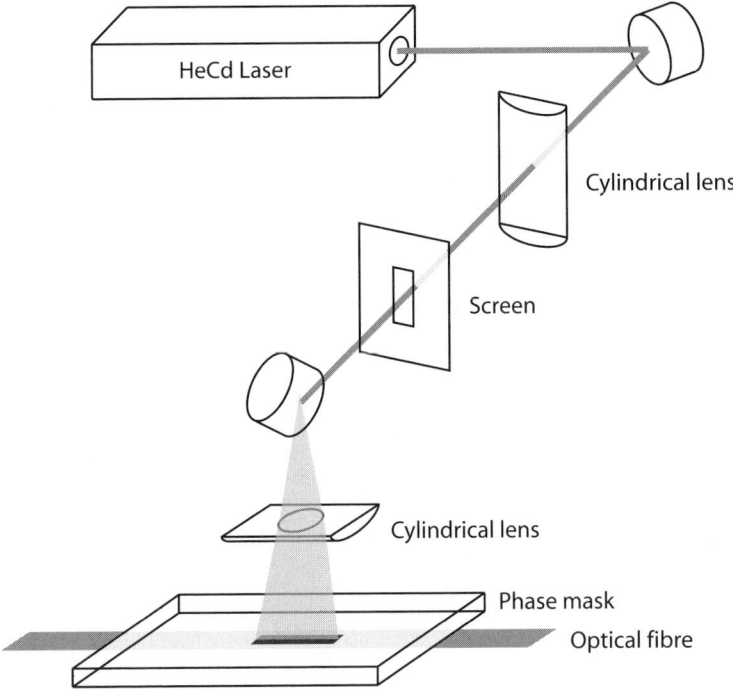

Fig. 10.3. The experimental set-up used to write Bragg gratings in mPOF. After Dobb et al. [2005].

et al. [2005] found that gratings could not be written unless the writing beam was incident on the "flat" side of the hexagonal structure. This requirement has not been reported in the literature on FBGs in silica MOFs, although the refractive index of the two materials and the relative size of the UV wavelength and microstructure are approximately the same. The alignment of the fibres is less critical when the fibre is properly supported, to prevent it bowing downwards.

The fibres used in the initial experiments included single-moded, few-moded and multimoded fibres. None of the fibres was doped with photosensitive material. Grating growth was monitored in reflection during the writing process. They appeared after approximately 20 minutes of exposure and saturated after 1 hour.

The spectra for these three fibres are shown in Fig. 10.5. The strongest grating to date (for the few-moded fibre) reflected about 50% of the incident light. An unexpected result is that the increase in core size did not cause a large increase in the spectral width of the grating. The fifteen-fold increase in core diameter between the single-mode and multimode fibre tested (3.9 μm

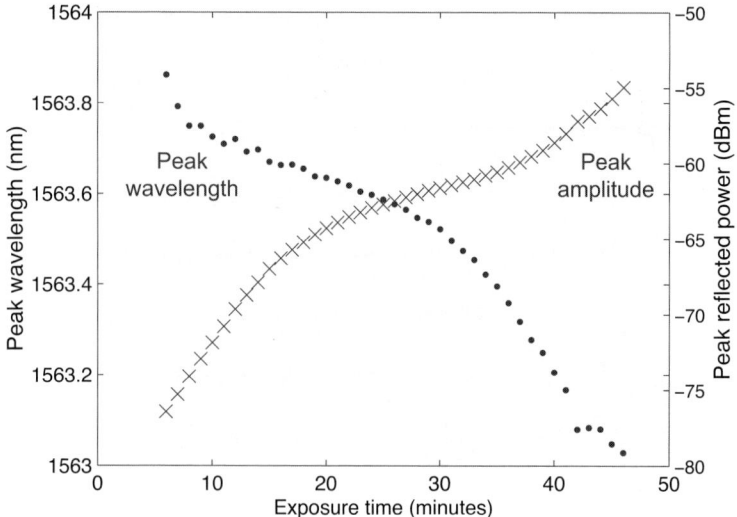

Fig. 10.4. The time evolution of gratings during the writing process. The shift in wavelength is caused by the progressive change in average refractive index. *Data courtesy of Aston University.*

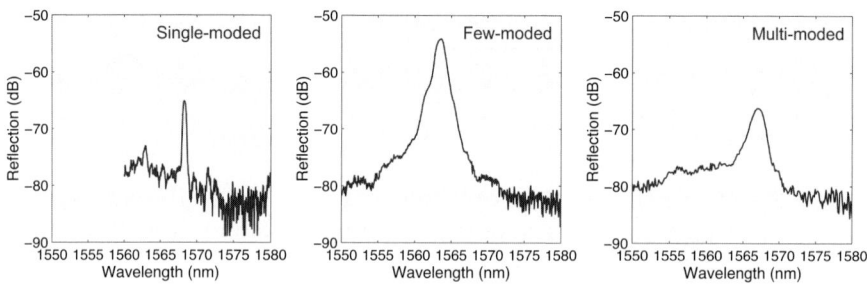

Fig. 10.5. Reflection spectra for gratings written in the single-mode, few-moded and multimode fibres. *Data courtesy of Aston University.*

compared to 60 μm) resulted in only a four-fold increase in grating width (FWHM of 0.5 nm and 2 nm respectively) with only a single peak.

The reason for this result was revealed by subsequent modelling (Fig. 10.6). The multimode fibre has only a small number of low-loss modes which are close in in their effective indices. Thus the loss of the higher order modes is too large to allow them to be observed, and the density of observable modes is too high for the individual peaks to be resolved.

This result is significant for several reasons. Large core fibres allow cheap, broad spectrum LEDs to be used as a source, rather than the more expensive single spatial mode emitters needed for use with single-mode fibre. It is likely that the 3 nm spectral width of the multimode fibre grating will be sufficient for many applications, particularly in sensing where large wavelength shifts are to be expected. It also opens the prospect of tunable mPOF lasers. Tunable POF gratings have already been used in conjunction with erbium-doped fibre [Liu et al. 2006], and lasing has been obtained in POF doped with laser dyes (see Chapter 11 and [Kuriki et al. 2000, Argyros et al. 2004]). In POF with small cores however, stimulated Raman scattering depletes the pump and prevents lasing. POF lasers must therefore be made with large core fibres. The multimode grating results show that the requirement for a large core is compatible with using a FBG.

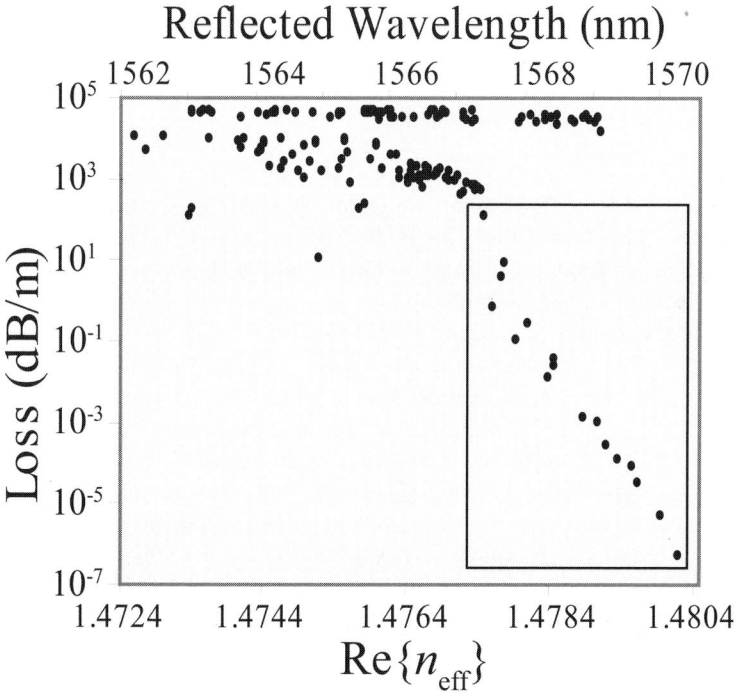

Fig. 10.6. Dependence of confinement loss on $\mathrm{Re}\{n_{\mathrm{eff}}\}$, showing that low-loss modes occur over a small range (highlighted). This gives rise to the relatively narrow reflection peak observed in the multimode fibre.

The strain-tuning behaviour of the gratings was also investigated. The few-moded mPOF was glued to translation stages with epoxy and stretched, and the reflectance spectrum measured. The results are shown in Fig. 10.7.

The resulting tuning range of 41 nm should be considered as a lower bound, as the measurements were limited by the spectral width of the source and is lower than reported previously for PMMA fibres [Xiong et al. 1999]. The tuning was reversible below the elastic limit, however, the reflection peak did

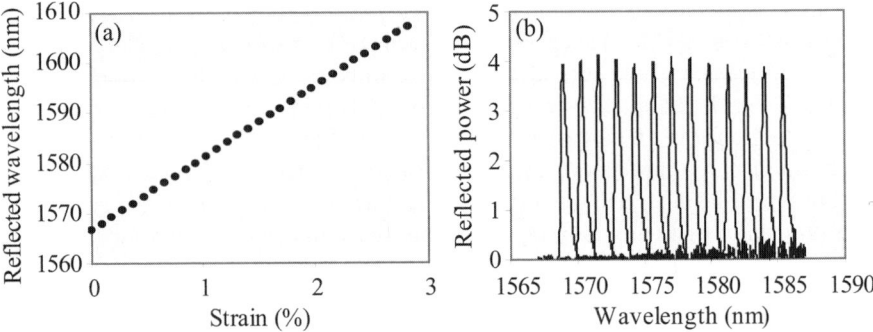

Fig. 10.7. (a) Reflected wavelength of a grating in a few-moded mPOF as a function of strain. (b) Reflection spectra of the grating for a selected range of strains showing little effect on the reflected power. Measurements provided by Mark Hiscocks, University of New South Wales, Australia.

not return to its original value immediately when going from large strains back to zero. The elastic limit for PMMA fibres has been reported at around 10% [Dobb et al. 2006a], but may be influenced by the processing history and molecular weight of the polymer.

Challenges remain in making FBGs in mPOF. The strength of the gratings needs to be increased, and they should be written for the visible region where the polymers are most transparent. An attempt to increase the strength of the grating by adding a photosensitive dopant to the core (*trans*-4-stilbenemethanol) increased the speed of inscription but not the grating strength [Argyros 2006]. Another possibility would be to modify the inscription system so as to reduce the effect of the holes around the core. Hill gratings may be one possible approach here. Given the likely applications for FBGs, and the possibility of using them with large core fibres, there is a strong incentive for improvement in performance.

10.3 Long Period Gratings

10.3.1 LPG In Microstructured Fibres

LPGs have been successfully made in MOFs by a variety of methods including UV inscription [Eggleton et al. 2000], a CO_2 laser [Kakarantzas et al. 2002, Zhu et al. 2003], arc discharge [Morishita and Miyake 2004, Humbert et al. 2004, Dobb et al. 2006b], acoustic waves [Dioz et al. 2000], mechanical pressure [Lim et al. 2004] and using the infusion of microfluidic plugs into the holes [Kerbage and Eggleton 2003].

In conventional fibres, LPGs have been far less investigated than FBGs. In MOFs, the reverse is true. This may be due to the greater difficulty in writing

FBGs in MOFs, but it may also be because MOFs offer particular benefits to LPGs. Variations to the microstructure offer enormous scope to optimize the properties of the cladding modes. The microstructure allows other materials to be introduced into the fibre, and LPGs allow these materials to affect the transmission properties in the core.

Most of the LPG applications explored in MOFs use one or other of these advantages. Using the microstructure to control waveguide properties, fibre sensors and filters have been made that are largely invariant to temperature [Zhu et al. 2003, Dobb et al. 2006b] and potentially other perturbations. Asymmetric arrangements of holes have been found to give directional bend sensitivity [Dobb et al. 2006b], while Westbrook et al. [2000] used the holes of a silica MOF to introduce polymer, and used the high temperature sensitivity of the polymer to give thermal tuning of the grating properties. An identical pair of MOF LPGs has also been used to produce an all fibre Mach-Zehnder interferometer [Lim et al. 2004].More recently, MOF LPGs have been used for label-free detection of biomolecules. Biomolecules immobilized on the surface of the cladding holes were found to give a measurable shift in the resonant frequency of the grating. This system was able to detect layer thicknesses within a few nm [Rindorf et al. 2006].

10.3.2 LPGs In Polymer Fibres

LPGs are relatively new in polymer. However, a permanent LPG has been created in a POF [Li et al. 2005b] using a 275 µm amplitude mask and a mercury lamp. This resulted in a grating depth of 3 dB.

Temporary LPGs in microstructured polymer fibres

Temporary LPGs in mPOF were created using template imprinting [van Eijkelenborg et al. 2004], following an earlier method [Digonnet et al. 2000]. The experimental arrangement used is shown in Fig. 10.8 but without the heated stage. A PMMA rod with 150 triangular grooves of 0.2 mm depth and a period of 1 mm was pressed onto the fibre over a length of 15 cm. The same technique has also been used for silica MOFs [Lim et al. 2004], including those with for hollow cores [Steinvurzel et al. 2006]. This simple technique can be simply varied to alter the performance of the grating. The grating period can be varied by changing the angle between the fibre and the grooves, and it is easy to change the length of the grating and the applied pressure. An additional advantage for microstructured fibres is that the previously noted sensitivity to the orientation of the microstructure [Dobb et al. 2005] in UV inscription does not apply.

The sensitivity of the LPG spectra to the applied load pressure was measured. The results are shown in Fig. 10.9. The centre wavelength of the resonance at 577 nm was found to shift by 1.7 nm/kg due to the average index change induced by the pressure (with similar values for the other resonances).

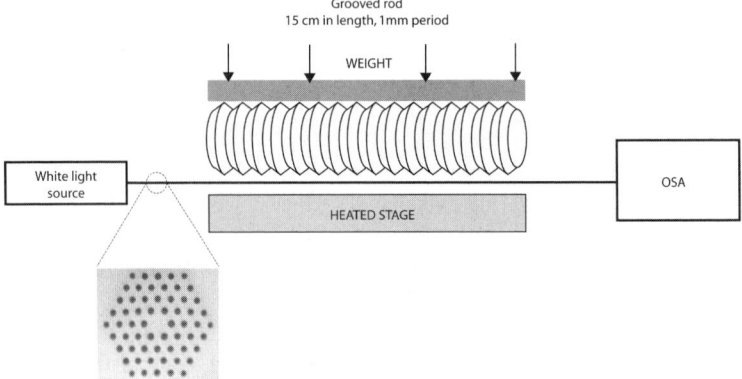

Fig. 10.8. Experimental arrangement for the mechanical imprinting of LPGs in mPOF

The strength of this resonance was found to grow on average by 4.8 dB/kg and returned to zero on the removal of the pressure. These results were found to be affected by the fibre fabrication conditions, in particular the tension employed during the draw.

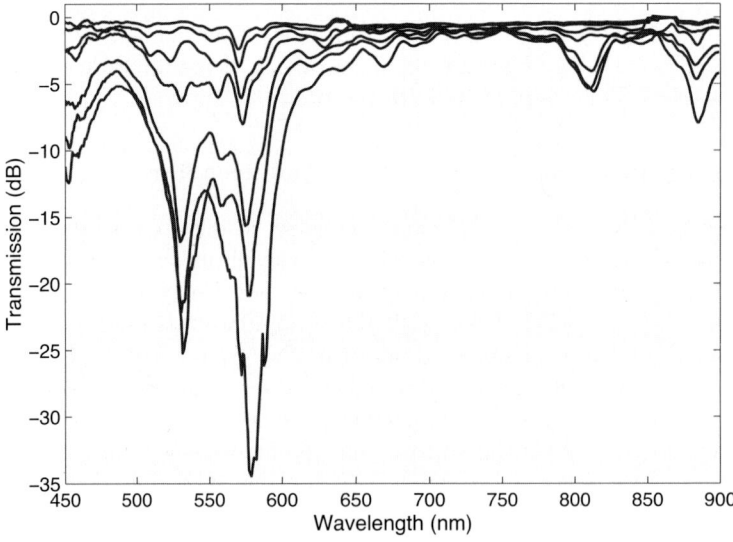

Fig. 10.9. Transmission spectra of a mechanically induced LPG of 15cm length in mPOF while increasing the applied weight from 0.33kg (top curve) to 0.65, 1.75, 2.75, 3.75, 4.75 and 5.85 kg (bottom curve). After van Eijkelenborg et al. [2004].

Stable LPGs in microstructured polymer fibres

A modification of this imprinting technique was used to produce stable LPGs in mPOF. A heated base (shown in Fig. 10.8) was used to apply heat during the imprinting process to allow permanent (or at least very long term) deformation of the fibre [Hiscocks et al. 2006]. The rest of the experimental set-up was unchanged. Variations of the heat imprinting method were used to determine the best parameters for inscribing gratings. This included varying the temperature of the heated base, the heating conditions (cyclic or continuous heating) and the weight applied on the grooved PMMA template. It was expected that the temperature required to write a stable grating would be above the glass transition point for PMMA (115 °C). It was found however that lower temperatures were optimal and again depended on the fibre draw conditions. The best gratings were written by keeping the temperature at around 60 - 80 °C. At approximately 60 °C a deepening of the attenuation peaks was observed, at which point the heated base was turned off. The heating was resumed when the spectrum was stable and the process was repeated until the desired attenuation was reached or the features developed no further. At this point the weight and grooved rod were removed and it was confirmed that the LPG features remained. There was an immediate reduction in the strength of the grating when the template was removed. The complete writing process took approximately 10 - 20 minutes. The best results were obtained with approximately 1 - 3 kg distributed along the template (approximately 60 - 200 g/cm).

Gratings imprinted by this method were repeatable, with similar writing conditions resulting in features with consistent locations. The main variation experienced was the depth of certain features. Shifting the position of the weight along the grating or changing the temperature at which the grating was written enhanced or suppressed certain features. An example spectrum is shown in Fig. 10.10.

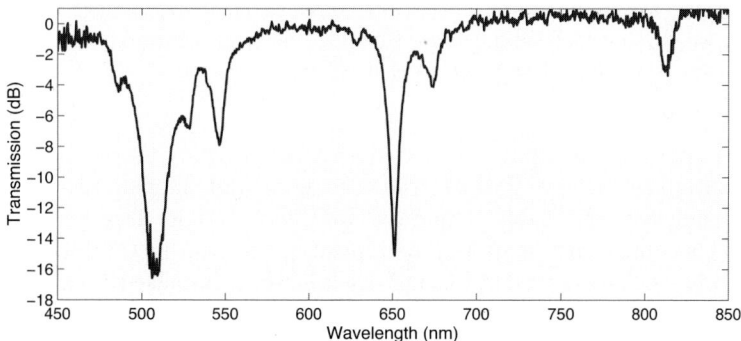

Fig. 10.10. Transmission spectrum of a LPG inscribed by heat imprinting with template of period 1 mm. After Hiscocks et al. [2006].

Accelerated aging tests were used to assess the long-term stability of these gratings. Under raised temperature conditions the gratings experienced a drop in grating strength over a period of several minutes. After this initial decrease, the grating strength was found to be quite stable. This is attributed to the grating having both a material (stress) and a waveguide component due to physical deformation. Low temperature and high pressure, which would produce the strongest stress component, produced the least stable grating. The results of the aging tests are shown in Fig. 10.11. Tests were also conducted over longer periods. A grating made under similar conditions to that labelled LWHT was stored at 60 °C for a period of 2 weeks and maintained a grating strength of approximately 8.7 dB.

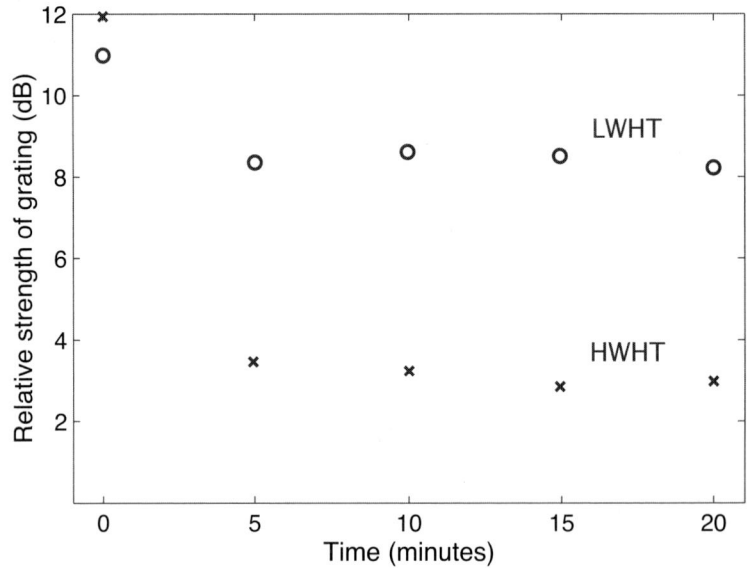

Fig. 10.11. Grating strength versus time stored at 60 °C. One LPG was written with low weight and high temperature (LWHT) the other with high weight and low temperature (HWLT). After Hiscocks et al. [2006].

The experimental results were compared to theory by using Eq. 10.2, and determining the effective index of the core mode and the first three cladding modes using the adjustable boundary condition method [Issa and Poladian 2003]. This confirmed the approximate positions for the spectral features and general trends. The agreement was not exact however because the experimental structure differed slightly from the idealised structure that was modelled.

The sensitivity of LPGs to changes to a small number of holes in the cladding structure has also been shown to be quite significant [Dobb et al. 2006b]. An indication of the effect of systematically varying the microstructure

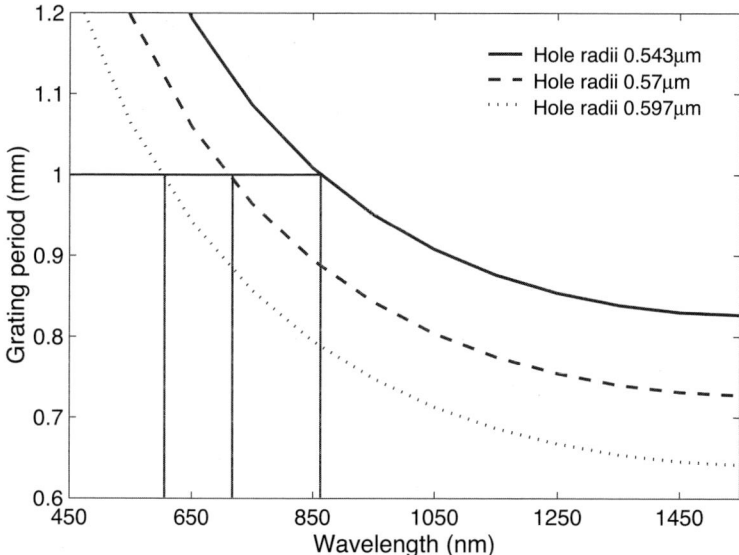

Fig. 10.12. Calculated resonances for first-order coupling between the core and first cladding modes for varying hole size. This is indicative of changes experienced by all resonances. After Hiscocks et al. [2006].

on the cladding stucture is shown in Fig. 10.12. Increasing the size of all holes in the microstructure by 10% causes a large shift in the predicted resonance, in this case a shift of approximately 265 nm for a grating period of 1 mm. Thus the sensitivity of LPGs to variations in the microstructure remains a double-edged sword. It has enormous potential sensitivity, but at the same time places strong requirements on the uniformity of the fibre fabrication. The use of LPGs in mPOF none the less presents enormous opportunities which are only just beginning to be explored.

10.4 Challenges And Future Directions

Probably the major challenge facing the application of gratings to microstructured fibres is the difficulty in writing them. The cladding microstructure inevitably scatters the write-beam, and reduces the efficiency of the writing process. The most promising approach to resolving this may well be in using new fibre designs, which allow a more direct access to the core. A further issues with polymer gratings is stability, particularly at elevated temperatures, and obtaining gratings in the that operate in the transparency region of the polymer.

There are however huge potential benefits if these challenges are met. They include using the gratings to manipulate the dispersion properties of the fibre, to obtain for example better nonlinear effects, as well as sensing applications. The latter is particularly significant in the context of aqueous sensing, also enabled by the use of microstructured fibres. Long period gratings offer particular opportunities, because of the ability to use changes in the cladding microstructure to tune the properties of the cladding modes.

References

Allsop, T, Zhang, L, Webb, D J, and Bennion, I (2002). Discrimination between temperature and strain effects using first and second-order diffraction from a long-period grating. *Optics Communications*, 211(1-6):103–8.

Argyros, A (2006). *Bragg Reflection and Bandgaps in Microstructured Optical Fibres*. PhD dissertation, School of Physics, The University of Sydney, Sydney, Australia.

Argyros, A, van Eijkelenborg, M A, Jackson, S D, and Mildren, R P (2004). A microstructured polymer fiber laser. *Optics Letters*, 29(16):1882–4. see also [Argyros et al. 2005].

Argyros, A, van Eijkelenborg, M A, Jackson, S D, and Mildren, R P (2005). Reply to comment on "Microstructured polymer fiber laser". *Optics Letters*, 30(14):1829–30.

Beugin, V, Bigot, L, Niay, P, Lancry, M, Quiquempois, Y, Douay, M, Mélin, G, Fleureau, A, Lempereur, S, and Gasca, L (2006). Efficient Bragg gratings in phosphosilicate and germanosilicate photonic crystal fiber. *Applied Optics*, 45(32):8186–93.

Caning, J, Groothoff, N, Buckley, E, Ryan, T, Lyytikainen, K, and Digweed, J (2003). All-fibre photonic crystal distributed Bragg reflector (pc-dbr) fibre laser. *Optics Express*, 11(17):1995–2000.

Choi, S, Eom, T J, Yu, J W, Lee, B H, and Oh, K (2002). Novel all-fiber bandpass filter based on hollow optical fiber. *IEEE Photonics Technology Letters*, 14(12):1701–3.

Daxhelet, X and Kulishov, M (2003). Theory and practice of long-period gratings:when a loss becomes a gain. *Optics Letters*, 28(9):686–8.

Digonnet, S Savin M J F, Kino, G S, and Shaw, H J (2000). Tunable mechanically induced long-period fiber gratings. *Optics Letters*, 25(10):710–2.

Dioz, A, Birks, T A, Reeves, W H, Mangan, B J, and Russell, P St J (2000). Excitation of cladding modes in photonic crystal fibers by flexural acoustic waves. *Optics Letters*, 25(20):1499–1501.

Dobb, H, Carroll, K, Webb, D J, Kalli, K, Komodromos, M, Themistos, C, Peng, G D, Argyros, A, Large, M C J, van Eijkelenborg, M A, Fang, Q, and Boyd, I W (2006a). Grating based devices in polymer optical fibre. In *Proceedings of the SPIE Photonics Europe Conference*, Strasbourg, France.

Dobb, H, Kalli, K, and Webb, D J (2006b). Measured sensitivity of arc-induced long-period grating sensors in photonic crystal fibre. *Optics Communications*, 260(1):184–91.

Dobb, H, Webb, D J, Kalli, K, Argyros, A, Large, M C J, and van Eijkelenborg, M A (2005). Continuous wave ultraviolet light-induced fibre Bragg gratings in few- and single-moded microstructured polymer optical fibres. *Optics Letters*, 30(24):3296–8.

Eggleton, B J, Kerbage, C, Westbrook, P S, Windeler, R S, and Hale, A (2001). Microstructured optical fiber devices. *Optics Express*, 9(13):698–713.

Eggleton, B J, Westbrook, P S, White, C A, Kerbage, C, Windeler, R S, and Burdge, G L (2000). Cladding mode resonances in air-silica microstructure optical fibers. *Journal of Lightwave Technology*, 18(8):1084–100.

Eggleton, B J, Westbrook, P S, Windeler, R S, Spalter, S, and Strasser, T A (1999). Grating resonances in air-silica microstructured optical fibers. *Optics Letters*, 24(21):1460–2.

Frazão, O, Carvalho, J P, Ferreira, L A, Araújo, F M, and Santos, J L (2005). Discrimination of strain and temperature using Bragg gratings in microstructured and standard optical fibres. *Measurement Science and Technology*, 16:2109–13.

Fu, L B, Marshall, G D, Bolger, J A, Steinvurzel, P, Magi, E C, Withford, M J, and Eggleton, B J (2005). Femtosecond laser writing Bragg gratings in pure silica photonic crystal fibres. *Electronics Letters*, 41(11):638–40.

Groothoff, N, Canning, J, Buckley, E, Lyttikainen, K, and Zagari, J (2003). Bragg gratings in air-silica structured fibers. *Optics Letters*, 28(4):233–5.

Han, Y-G, Lee, S B, Jin, C-S K J, Kang, U, Paek, U-C, and Chung, Y (2003). Simultaneous measurement of temperature and strain using dual long-period fiber gratings with controlled temperature and strain sensitivities. *Optics Express*, 11(5):476–81.

Hill, K O, Fujii, Y, Johnson, D C, and Kawasaki, B S (1978). Photosensitivity in optical fibre waveguides: Application to rejection filter application. *Applied Physics Letters*, 32(10):647–9.

Hiscocks, M P, van Eijkelenborg, M A, Argyros, A, and Large, M C J (2006). Stable imprinting of long-period gratings in microstructured polymer optical fibre. *Optics Express*, 14(11):4644–9.

Humbert, G, Malki, A, Fevrier, S, Roy, P, and Pagnoux, D (2004). Characterizations at high temperatures of long period gratings written in germanium-free air-silica microstructure fiber. *Optics Letters*, 29(1):38–40.

Issa, N A and Poladian, L (2003). Vector wave expansion method for leaky modes of microstructured optical fibres. *Journal of Lightwave Technology*, 21(4):1005–12.

Kakarantzas, G, Birks, T A, and Russell, P St J (2002). Structural long-period gratings in photonic crystal fibers. *Optics Letters*, 27(12):1013–15.

Kashyap, R (1999). *Fiber Bragg Gratings*. Academic Press, San Diego, USA.

Kerbage, C and Eggleton, B J (2003). Tunable microfluidic optical fiber gratings. *Applied Physics Letters*, 82(9):1338–40.

Kuriki, K, Kobayashi, T, Imai, N, Tamura, T, Koike, Y, and Okamoto, Y (2000). Organic dye-doped polymer optical fiber lasers. *Polymers for Advanced Technologies*, 11:612–6.

Lee, K S and Erdogan, T (2001). Fiber mode conversion with tilted gratings in an optical fiber. *Journal of the Optical Society of America A*, 18(5):1176–85.

Li, Y F, Salisbury, F C, Zhu, Z M, Brown, T G, Westbrook, P S, Feder, K S, and Windeler, R S (2005a). Interaction of supercontinuum and Raman solitons with microstructure fiber gratings. *Optics Express*, 13(3):998–1007.

Li, Z, Tam, H Y, Xu, L, and Zhang, Q (2005b). Fabrication of long-period gratings in poly(methyl methacrylateco-methyl vinyl ketone-co-benzyl methacrylate)-core polymer optical fiber by use of a mercury lamp. *Optics Letters*, 30(10):1117–9.

Lim, J H, Lee, K S, Kim, J C, and Lee, B H (2004). Tunable fiber gratings fabricated in photonic crystal fiber by use of mechanical pressure. *Optics Letters*, 29(4):331–3.

Liu, H B, Liu, H Y, Peng, G D, and Chu, P L (2004). Novel growth behaviors of fiber Bragg gratings in polymer optical fiber under UV irradiation with low power. *IEEE Photonics Technology Letters*, 16(1):159–61.

Liu, H Y, Liu, H B, and Peng, G D (2006). Polymer optical fibre Bragg gratings based fibre laser. *Optics Communications*, 266(1):132–5.

Liu, H Y, Liu, H B, Peng, G D, and Chu, P L (2003). Observation of type I and type II gratings behavior in polymer optical fiber. *Optics Communications*, 220(4-6):337–43.

Liu, H Y, Peng, G D, and Chu, P L (2001). Thermal tunability of polymer optical fibre Bragg gratings. *IEEE Photonics Technology Letters*, 13(8):824–6.

Liu, Y, Zhang, L, and Bennion, I (1999). Fibre optic load sensors with high transverse strain sensitivity based on long-period gratings in b/ge co-doped fibre. *Electronics Letters*, 35(8):661–3.

Morishita, K and Miyake, Y (2004). Fabrication and resonance wavelengths of long-period gratings written in a pure-silica photonic crystal fiber by the glass structure change. *Journal of Lightwave Technology*, 22(2):625–30.

Othonos, A and Kalli, K (1999). *Fiber Bragg gratings: Fundamentals and applications in telecommunications and sensing.* Artech House Publishers.

Rindorf, L, Jensen, J B, Dufva, M, Pedersen, L H, Høiby, P E, and Bang, O (2006). Photonic crystal fiber long-period gratings for biochemical sensing. *Optics Express*, 14(18):8224–31.

Rocha, M L, Borin, F, Monteiro, H C L, Horiuchi, M R, de Barros, M R X, Santos, M A D, Oliveira, F L, and Oes, F D S (2005). Mechanical tuning of fiber Bragg grating for optical network applications. *Journal of Microwave and Optoelectronics*, 4(1):1–11.

Søndergaard, T (2000). Photonic crystal distributed feedback fiber lasers with Bragg gratings. *Journal of Lightwave Technology*, 18(4):589–97.

Stegall, D B and Erdogan, T (2000). Dispersion control with use of long-period gratings. *Journal of the Optical Society of America A*, 17(2):304–12.

Steinvurzel, P, Moore, E D, Mägi, E C, Kuhlmey, B T, and Eggleton, B J (2006). Long period grating resonances in photonic bandgap fiber. *Optics Express*, 14:3007–14.

Tomlinson, W J, Kaminow, I P, Fork, E A Chandross R L, and Silfvast, W T (1970). Photoinduced refractive index increase in poly(methylmethacrylate) and its applications. *Applied Physics Letters*, 16(12):486–9.

van Eijkelenborg, M A, Padden, W, and Besley, J A (2004). Mechanically induced long-period gratings in microstructured polymer fibre. *Optics Communications*, 236:75–8.

Vengsarkar, A M, Lemaire, P J, Judkins, J B, Bhatia, V, Erdogan, T, and Sipe, J E (1996). Long-period fiber gratings as band-rejection filters. *Journal of Lightwave Technology*, 14(1):58–64.

Webb, D J, Aressy, M, Argyros, A, Barton, J S, Dobb, H, van Eijkelenborg, M A, Fender, A, Jones, D C, Kalli, K, Kukureka, S, Large, M C J, MacPherson, W, Peng, G D, and Silva-Lopez, M (2005). Grating and interferometric devices in POF. In *Proceedings of the International Conference on Polymer Optical Fibre*, volume 14, pages 325–8, Hong Kong, China. Session XIV 'Polymer Optical Devices'.

Westbrook, P S, Eggleton, B J, Windeler, R S, Hale, A, Strasser, T A, and Burdge, G L (2000). Cladding-mode resonances in hybrid polymer-silica microstructured optical fiber gratings. *IEEE Photonics Technology Letters*, 12(5):495–7.

Xiong, Z, Peng, G D, Wu, B, and Chu, P L (1999). 73 nm waveguide tuning in polymer optical fiber Bragg gratings. In *Proceedings of the Australian Conference on Optical Fibre Technology*, pages 2–5, Sydney, Australia.

Yu, J M, Tao, X M, and Tam, H Y (2004). *Trans*-4-stilbenemethanol-doped photosensitive polymer fibers and gratings. *Optics Letters*, 29(2):156–8.

Zhu, Y, Shum, P, Chong, H-J, Rao, M, and Lu, C (2003). Strong resonance and highly compact long-period grating in a large-mode-area photonic crystal fiber. *Optics Express*, 11(16):1900–5.

Material Additives for Microstructured Polymer Optical Fibres

Eye of newt and toe of frog, Wool of bat and tongue of dog,
Adders fork and blind-worms sting, Lizards leg and owlets wing,
For a charm of powerful trouble, Like a hell-broth boil and bubble.

William Shakespeare, The Scottish Play

Two doping methods that specifically use the holey nature of mPOF are discussed. Dopants or particles can be introduced into the holes of the preform or the cane. Both methods are described here, and results showing the fluorescence from fibres doped with organic dyes, embedded organo-silica nanoparticles and quantum dots are presented. Indeed, an mPOF laser has been demonstrated using the solution doping method.

11.1 Introduction

The doping of conventional silica fibres is a long-established technique underpinning much of today's long-haul telecommunications technology. A silica preform is doped (e.g. using a high temperature process such as modified chemical vapour deposition) to produce regions with slightly different refractive indices that on drawing become the core and cladding of the fibre. In addition to the co-deposition (with silica) of a range of oxides such as GeO_2, P_2O_5 and B_2O_3 on the wall of the substrate tube, rare-earth doping is commonly used to obtain fluorescence or amplification in silica fibres.

In silica fibres, however, the potential to modify fibre properties by doping is limited (i) by the high processing temperatures required (up to 2000 °C) which would cause many materials to decompose (essentially all organic materials decompose by 400 °C), and (ii) by material compatibility issues. Certainly conventional glass fibres do not allow a great deal of flexibility as too much dopant can lead to induced stress as the fibre cools to room temperature and/or phase separation leading to non-homogeneities within the fibre.

Doping in polymer optical fibre (POF) is only used for specialty fibre. Simple step-index POF is made by drawing an undoped preform with a low refractive index photo-curable coating as a cladding [Daum et al. 2002]. However for the fabrication of graded-index polymer fibre (GIPOF) for high bandwidth applications, a dopant is introduced to create a parabolic refractive index profile that provides the guiding core [Ishigure et al. 2000]. This can be achieved using a number of methods, such as polymerisation in a centrifuge [Shin et al. 2004].

A range of organic dopants have also been introduced into POF (including fluorescent or photosensitive materials such as Rhodamine dyes), which have traditionally been added at the polymerisation stage in preform manufacture [Naritomi et al. 2004]. Inorganic rare-earth ions have fluorescence properties that can be exploited in solid-state amplifiers and lasers functioning over the visible to mid-IR range [Digonnet 2001]. As metal ions are charged, they cannot be incorporated into polymers directly, requiring that they be stabilised through binding to organic ligands or incorporation into an organometallic compound. These organic ligands have been used to incorporate metal ions into POF, as with organic dopants by incorporation at the monomer stage prior to polymerisation into a preform [Kuriki et al. 2001].

The potential to modify conventional POF fibre properties by doping is limited. Organic dopants can be used and chelates allow doping to be extended to rare-earth ions, but in all cases the chemistry involved is far from trivial. Commercially reliable methods to produce graded-index POF (with a near parabolic dopant concentration distribution) have taken several decades to develop, but these merely provide light guidance within the fibre with no other functionality. Clearly dopant usage in POF has been neither simple nor amenable to generalisation.

With the mPOF doping techniques described in this chapter, the foundation is laid for making new fibres such as doped nonlinear mPOF for supercontinuum generation, photosensitive mPOF for grating writing, surface-coated mPOF with controlled-release nanoparticles for biosensing, fibres with embedded single-photon sources (using nitrogen-vacancy diamond nanoparticle inclusions), doped mPOF for blue and/or green lasers and magneto-optically active fibres for optical isolators.

It is possible to introduce dopants into the preform using the microstructure, rather than at the monomer stage as has been done for POF. This allows doped mPOFs to be made easily, using commercially available low-loss polymer. As shown previously [Zagari et al. 2004b], the polymer walls between the holes forming the microstructure in an mPOF can be made less than a half a micron thick. Such structures are sufficiently thin that they can be considered as dense membranes.

The development of a generic process by which a range of nanoparticles, such as organosilica nanoparticles (OSNPs) and Quantum Dots (QDs), can be embedded into mPOF further removes the conventional limitations on polymer fibre doping, making mPOFs a universal platform for the creation of a very broad range of active fibres. The embedding of OSNPs in optical fibres is possible as their fabrication uses sol-gel synthesis to encapsulate a variety of materials [Finnie et al. 2005]. Fibres with dopant species that are incompatible with the polymer material can now be fabricated while the low processing temperatures involved in the fabrication allow for the incorporation of organic species. The resulting spatial distribution of the particles in mPOF is fixed, as opposed to a varying spatial distribution which currently occurs in silica fibres [Meissner et al. 2005].

Nanoparticle incorporation promises the solution to a number of known doping issues, one of which is photobleaching. Also known as photo-induced oxidisation, this process severely limits the useful lifetime of organics for light-emitting purposes [Song et al. 1995]. Photobleaching can be eliminated by using a technique to encapsulate dye in a nanoparticle with a core/shell structure [Tan et al. 2004]. The encapsulation minimises the contact between the dopant and potential oxidising agents, giving the fluorescent dye a virtually infinite lifetime. At the same time this allows precise control over the amount of dye that is be embedded within the fibre. Similarly, QDs are a promising type of nanoparticle as they are inherently resistant to photobleaching and their emission wavelength can be tailored by varying the size of the dots used (see e.g. [Pang et al. 2005] or www.evidenttech.com).

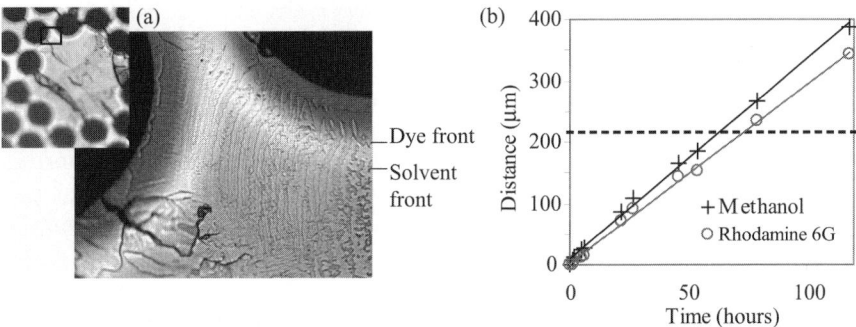

Fig. 11.1. (a) Cross-section of a preform during doping, showing two holes adjacent to the core with the dye and solvent fronts diffusing in from the holes (top). (b) Plot of the position of the dye and solvents fronts as a function of time. The linear dependence indicates Case II diffusion. The dotted line indicates the position of the core radius. After Large et al. [2004].

Introducing particles into an optical fibre can affect the transmission losses through enhanced scattering (see also Section 7.2.5). This can be kept to a minimum by (i) using particles that are significantly smaller than the wavelength transmitted (e.g. the 20 nm diameter particles as in Fig. 11.7, (ii) preventing clustering of particles which would act as scattering centers, (iii) matching the refractive index of the particle to that of the fibre material and (iv) operating at longer wavelengths.

11.2 Doping Methods

11.2.1 Solution Doping

At temperatures above the glass transition temperature, diffusion of molecules within a polymer can be described by Fick's laws [Neogi 1996]. In the glassy state however, deviations from Fickian behaviour occur because of the finite time required for the polymer molecules to accommodate the dopant molecules. When the dopant molecule mobility is much larger than the segmental relaxation, the diffusion is known as Case II diffusion. This is characterised by a sharply defined boundary between the swollen outer layer and the glassy interior of the polymer, and a diffusion front that advances with near uniform velocity. The dopant plasticises the polymer and the rate of penetration is proportional to the concentration at the diffusion front.

The transport of methanol in PMMA has been studied (see e.g. [Thomas and Windle 1978, 1980], including methanol/Rhodamine mixtures [Muto and Tahjika 1986]. These studies show that at ambient temperatures, PMMA exhibits Case II diffusion, a result consistent with our results (see Fig. 11.1), where the dye lags the solvent significantly, due to its larger molecular size. Note that the presence of dye molecules does not seem to affect the passage of the solvent through the polymer [Thomas and Windle 1978, Muto and Tahjika 1986]. It has been suggested that the mechanical deformation of the polymer brought about by osmotic swelling due to the solvent is the most important driver of Case II diffusion [Thomas and Windle 1980], with the dopant able to diffuse relatively quickly through the polymer behind the solvent front.

MPOFs are made using a two-stage draw technique, first producing cane (an intermediate preform) and then fibre. In our studies, doping was carried out after the first draw, at the cane stage when the holes are about 250 microns in diameter, sufficiently large to still allow solutions to pass through by attaching them to a syringe. The preforms were annealed before doping to alleviate any residual stress which could cause cracking when the solution was introduced, and to ensure that prior thermal history did not play a role in determining the dopant uptake. Rhodamine 6G (R6G) was dissolved in methanol, a good solvent for Rhodamine, but a non-solvent for PMMA. Methanol is also quite volatile, having a boiling point of 60 °C, which allows it to be readily removed from the polymer below its glass transition temperature of 115 °C. Other solvents would allow a wide variety of dopants to be introduced into PMMA.

Preforms were left in a solution of the dye/methanol until the diffusion fronts had met at the centre of the core region, a process that took up to three days at room temperature, depending on the preform design. They were

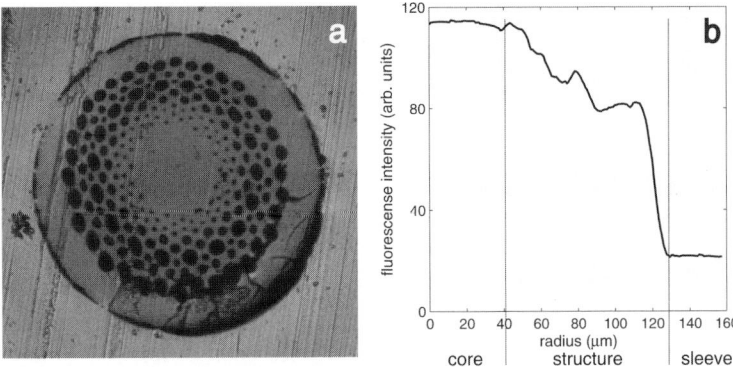

Fig. 11.2. (a) Cross-section of a doped mPOF. Orange coloured parts are Rhodamine doped. (b) Fluorescence intensity as a function of radius (averaged over all azimuthal angles) shows a uniform doping concentration in the core. After Large et al. [2004]. See also colour plate at front of book.

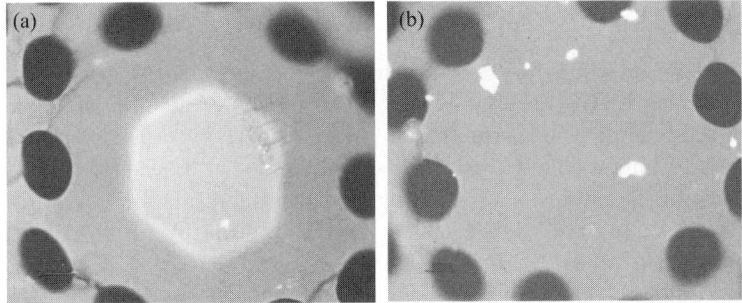

Fig. 11.3. Cross-section of a preform removed from the solution prior to the diffusion fronts meeting at the core (left) and the same preform after heating (right). The green is the fluorescence of the dye, viewed through an optical microscope using reflected light with the sample between crossed polarisers. After Large et al. [2004]. See also colour plate at front of book.

then dried to remove the methanol, by heating at 90 °C for several hours before being drawn to fibre. The removal of the solvent dramatically reduces dye diffusion. After its removal, there is no measurable change to the dye distribution, even when the preform is maintained at elevated temperatures for extended periods [Large et al. 2004]. The concentration of dye in the polymer can be varied by altering the dye concentration in the solution, the choice of solvent or the temperature. Using these techniques, uniformly doped samples with dopant concentrations ranging from 1 µmol/L to 1 mmol/L have been produced. An example of a fibre doped using this method is shown in Fig. 11.2.

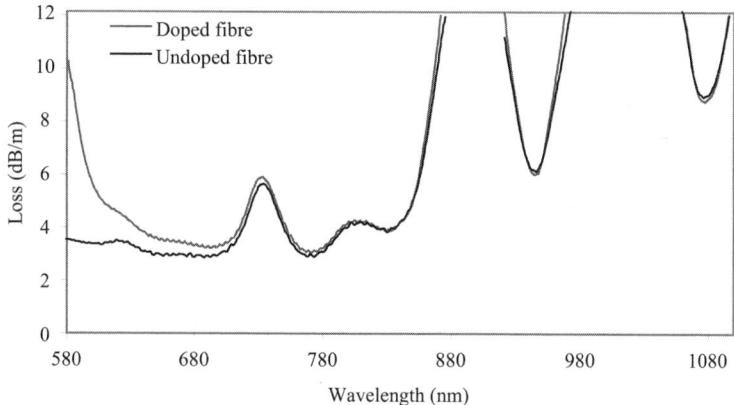

Fig. 11.4. Optical loss measurements of a doped and undoped fibre, drawn from the same preform, indicating that the doping does not introduce losses other than those due to the absorption of the dye (the sharp increase at the short wavelength region of the graph). After Large et al. [2004].

A preform can also be removed from the solution prior to the diffusion fronts meeting at the core, at which point a smaller amount of dopant has entered the polymer. Further diffusion during the drying stage however results in a uniform dopant concentration, as shown in Fig. 11.3. The resulting dopant concentration is lower than if soaking had been continued and the diffusion fronts had been allowed to meet. This offers an alternative route to controlling the dye concentration.

The complete removal of the solvent is clearly important. Thus thermogravimetric analysis (TGA) was carried out to determine whether solvent remained within the polymer matrix after the drying stage. The cane sample was heated at a rate of 2 °C/minute, and then held isothermally at 250 °C for 10 min. There was no evidence of enhanced weight loss at or around the boiling point of methanol, indicating that it had been successfully removed by the low temperature drying. Rather there was a small continuous weight loss of 1.49% over the course of the TGA run which was probably due to a small amount of high-temperature depolymerisation. Optical measurements (Fig. 11.4) also indicated that the doping process did not increase the attenuation of the resulting fibre, other than through the expected absorptions due the dye.

Figure 11.5(a) shows the absorption spectrum of the Rhodamine 6G dye in mPOF as determined by measuring the loss spectrum of a doped fibre and subtracting the loss of an undoped fibre. The two fluorescence spectra shown in this graph are for a very short fibre (a few mm) and a longer fibre (approximately 2 m). The spectral shift to the red as observed for the longer fibre results from emission and re-absorption of the fluorescence by the dye. This effect is illustrated by the photograph in (b). The fluorescence measurements were taken by launching light from an Argon-ion laser (514 nm - this laser line has been removed from the graph) into the core of the fibre. For a dye concentration around 1 mmol/L, the absorption peak reached approximately 20 dB/m.

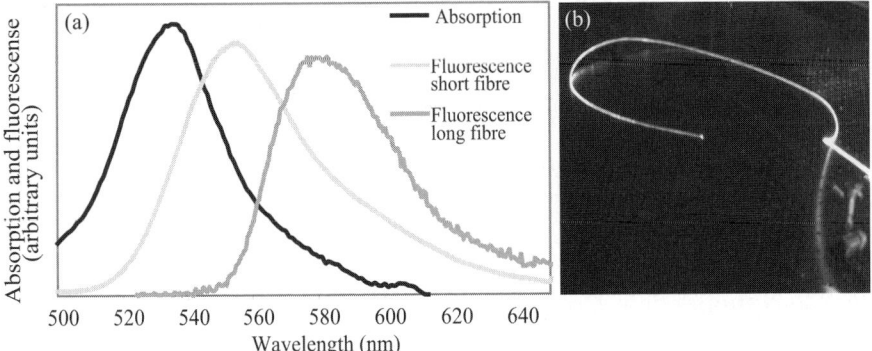

Fig. 11.5. (a) The absorption spectrum of the Rhodamine 6G dye and two fluorescence spectra corresponding to a very short fibre (a few mm) and a longer fibre (2 m). (b) Photograph illustrating the re-absorption process. After Large et al. [2004]. See also colour plate at front of book.

In an alternative approach, an acetone/Rhodamine solution was prepared and flushed through the preform for only 30 s, leaving a residual layer of dye which subsequently migrated through the preform when heated as shown in Fig. 11.6. Dye and solvent coat the surface of the holes and penetrate the polymer upon heating for 21 hours at 80 °C as shown in Fig. 11.3(a). Further diffusion is seem after an additional 1 hour (b) and 2 hours (c) of heating at 107 °C. A uniform concentration was achieved after the third hour. The sharp dopant front in Fig. 11.6(a) indicates that Case II diffusion occurs at the lower temperature, whilst Fickian diffusion occurs at the higher temperature, which is very close to the glass transition temperature of the polymer.

The doping process using Acetone is faster but more difficult to control than with Methanol. Also, as acetone is a solvent for PMMA, care must be taken to avoid damage to the microstructure. Another approach presented in the literature [Muto and Tahjika 1986] is to expose the PMMA to the solvent prior to immersion in the dye solution. This greatly increases the speed of the dye uptake into the pre-swollen polymer. The dye uptake dynamics of this system however are different to those of a 'dry' sample immersed in a dye/solvent mix.

It should be emphasised that the solution doping method is neither dye nor solvent specific. The low boiling points of acetone and methanol however made them ideal candidates for mPOF doping as they can be easily removed from the preforms, are readily available and easily handled.

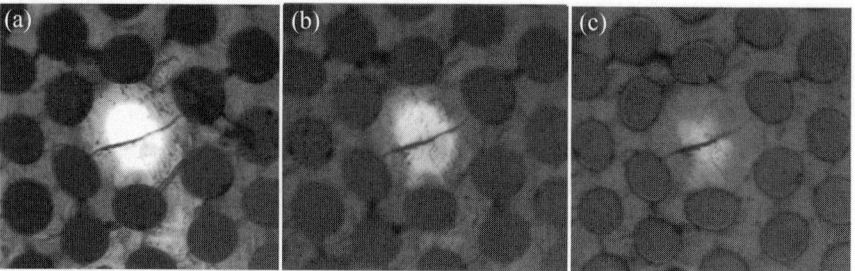

Fig. 11.6. Dopant diffusion through the core region using acetone as a solvent. Images taken after (a) 21 hours at 80 °Cand after an additional (b) 1 and (c) 2 hours at 107 °C.

11.2.2 Nanoparticle Doping

Recently advances in sol-gel technology have enabled the entrapment of organic molecules (including various therapeutic drugs) within nano-sized amorphous oxide matrices (see Fig. 11.7). The encapsulated species can then be released from these functional nanoparticles at a controlled rate. The first application of this technique focused on drug delivery for the treatment of cancer [Barbé and Bartlett 2001, Barbé et al. 2004, Finnie et al. 2005]].

Almost any organic molecule (and many inorganics) can be readily encapsulated within silica particles by combining sol-gel polymerisation with either spray-drying to produce micron-sized particles or emulsion chemistry to produce nanoparticles. The latter approach allows the production of particles with a homogeneous dopant distribution whilst employing ambient temperature processing (necessary for handling biologically active molecules and many organics). A critical feature of nanoparticles is that they tend to form micron-size aggregates that would scatter light, increasing optical losses, when incorporated into a fibre.

Two types of nanoparticles have been doped into mPOFs, organo-silica nanoparticles (OSNP) and quantum dots (QDs). OSNP particles of 300 nm diameter with encapsulated R6G were fabricated by sol-gel processing of phenyltrimethoxysilane. The second type of nanoparticles were "Hops Yellow" QDs composed of Cadmium Selenide/Zinc Sulfide semiconductor material with an amine terminal group synthetic surface coating (from Evident Technologies).

Both the OSNPs and QDs were suspended in de-ionised water. The OSNPs needed sonication to reduce aggregation. The suspended particles were then sucked up into the holes of a 5-hole suspended-core mPOF cane using a syringe, after which the cane was placed in an oven at 90 °C for four days until the water had evaporated and a layer of particles was left deposited on the walls of the holes. The cane was subsequently drawn to fibre in the usual way. The fibre cross section is shown in Fig. 6.9. This structure was chosen for initial trials because of its simplicity and its large hole size.

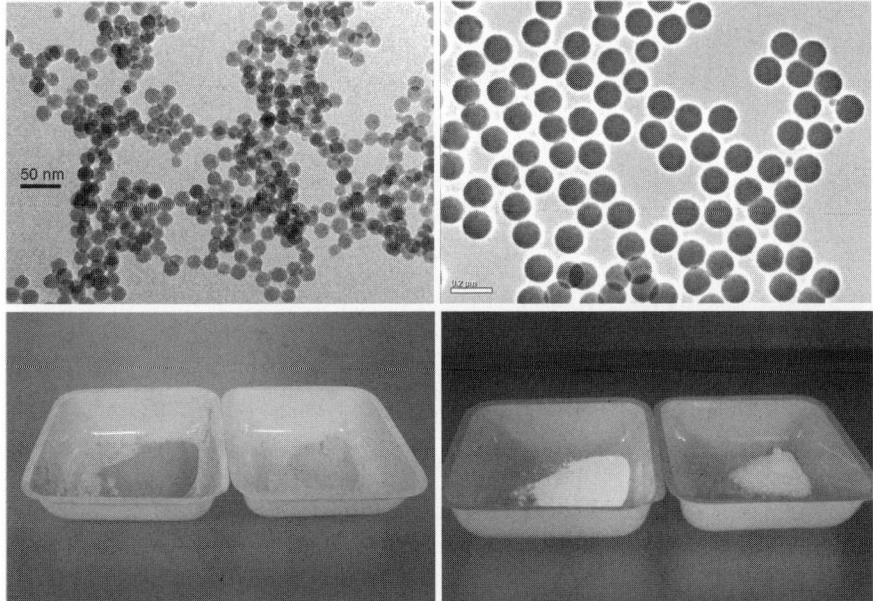

Fig. 11.7. Electron microscope images of (left) 20 nm and (right) 140 nm diameter silica nanoparticles containing R6G and Rhodamine B and their appearance under normal light and UV light *Images Courtesy of Dr. Chris Barbé of the Australian Nuclear Science and Technology Organisation (ANSTO).* See also colour plate at front of book.

11.3 Fluorescence Measurements

Four mPOFs with a 5-hole suspended-core structure similar to that shown in Fig. 6.9 and doped with (a) encapsulated R6G dye, (b) free R6G (using solution doping), (c) Hops Yellow QDs and (d) undoped (as reference) have been fabricated and characterised. These fibres had an outer diameter of 300 μm, 63 μm diameter holes and a core of 37.5 μm diameter. Fluorescence was observed by launching single-line Argon-ion laser light of 488 nm into the fibre core with a 10× microscope objective. This pump laser light was filtered out at the fibre output with a band stop rugate notch filter while the remaining light was coupled into a detection fibre connected to a spectrum analyser with a resolution of ±5 nm. The fluorescence emission spectra were taken both with the pump laser on and off (note that the laser was turned "off" by misaligning the back reflector of the Argon laser to be able to correct for fluorescence lines originating from the Argon-ion discharge). The resulting fluorescence spectra for all four fibres (a) to (d) together with the spectrum for "free" QDs [curve (e)] are shown in Fig. 11.8.

When embedded in the fibre, the fluorescence curves for free and encapsulated particles (either OSNP or QDs) exhibit shifts in peak fluorescence

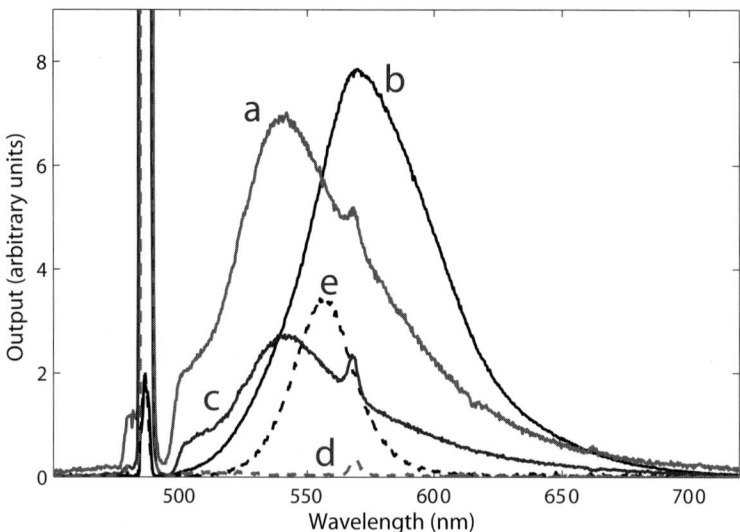

Fig. 11.8. Measured fluorescence curves for mPOF containing encapsulated R6G (a), free R6G in fibre (b), quantum dots in fibre (c), no dopant (d) and free quantum dots (e). All spectra show the remaining pump line at 488 nm and a Raman Peak at 569 nm. Relative intensities are not to scale. After Yu et al. [2006].

wavelength. There are two key reasons for this: (i) The environment surrounding the photon source, being either organo-silica or PMMA, shifts the electronic resonances, and (ii) the process of emission and re-absorption causes shifts dependent on the concentrations of the particles, as was illustrated by Fig. 11.5 [Kurian et al. 2002]. The latter is particularly important for the free R6G case, since solution doping leads to a high dopant concentration [Large et al. 2004, Argyros et al. 2004], and the resulting overlap of the emission and absorption bands for R6G result in a shifting of the maximum emission wavelength to the red.

11.4 An MPOF Amplifier And Laser

An obvious application of doping is to use the R6G laser dye to produce an mPOF amplifier and laser. To achieve this a 6 mm diameter cane with a structure as shown in Fig. 11.9 was doped using a saturated Rhodamine/acetone for 30 s, then dried by heating at 90 °C for 16 hrs. The final concentration of the dye in the PMMA core was estimated to be 1 mmol/kg.

The fibre was tested as an amplifier using the experimental setup shown in Fig. 11.10 by pumping the core at 532 nm using a frequency doubled Q-switched Nd:YAG laser operating at 10 Hz. A small signal (around 1 nJ per pulse) was provided by a conventional dye laser using Rhodamine dissolved in

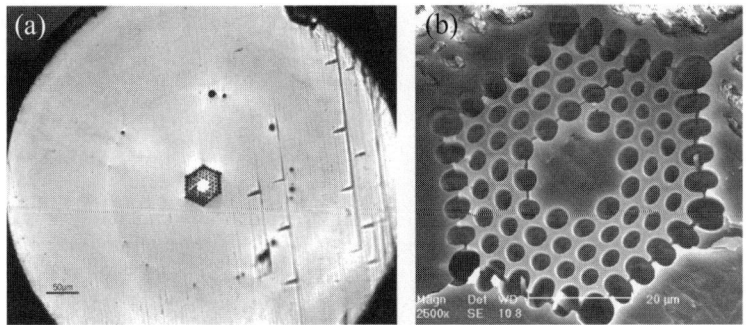

Fig. 11.9. (a) Optical microscope image of the mPOF used as an amplifier, and (b) a SEM image of the 18 μmdiameter core region. This fibre is 600 μmin diameter with 3.5 μmholes at 5.2 μmspacing. After Argyros et al. [2004].

methanol which could be tuned from approximately 560 to 585 nm. The pulses from the two lasers arrived simultaneously at the fibre with pulse lengths of 10 and 8 ns for the Nd:YAG and dye laser, respectively. The output was measured using a photodiode and the spectrum measured with a monochromator and a photomultiplier.

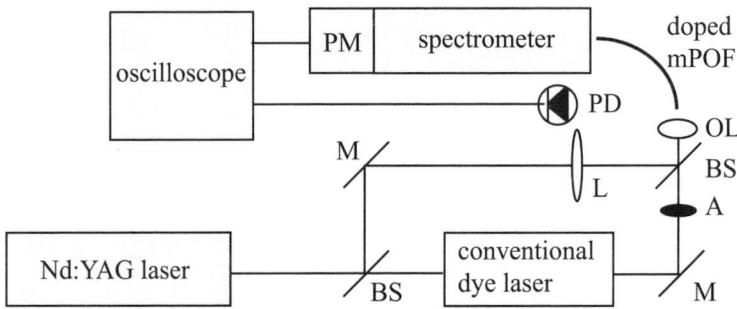

Fig. 11.10. Schematic of the setup used in the gain measurements. BS indicates a beamsplitter; M, mirror; A, attenuator; L, lens; OL, objective lens; PM, photomultiplier; PD, photodiode. The latter was used to trigger the oscilloscope. After Argyros et al. [2004].

Using a 2 m length of doped mPOF, the signal (dye laser) was tuned through the 560-585 nm range and the gain profile determined using a moderate pump pulse energy (about 50 μJ). The maximum gain occurred at 574 nm as shown in Fig. 11.11(a). The measured gain at 574 nm increased with pump power as shown in Fig. 11.11(b). The maximum gain observed was 30.3 dB (a factor of 1,072), requiring a launched pump energy of 325 μJ per shot (peak power of 32.5 kW). The gain saturates around this value while the low efficiency of 0.3% is due to a significant amount of amplified spontaneous emission being produced, caused by the large emission cross-section of the dye, the high peak pump power and to a lesser extent the numerical aperture of the fibre ($NA = 0.19$). A higher signal power could be expected to increase the efficiency as it would allow the signal, rather than the spontaneous emission, to dominate the stimulated emission.

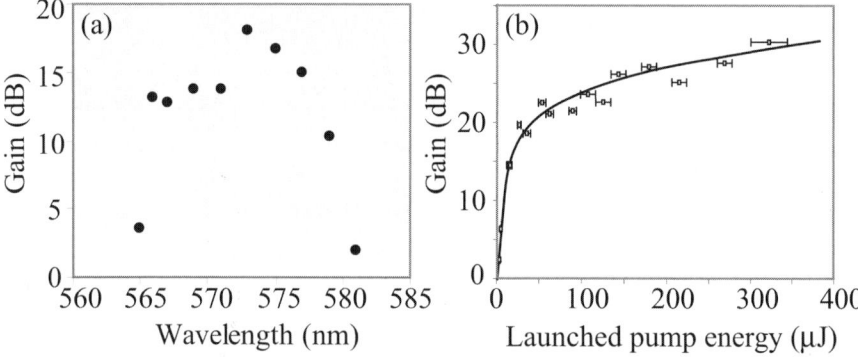

Fig. 11.11. (a) Gain as a function of wavelength, and (b) Gain at $\lambda = 574$ nm as a function of the launched pump energy. After Argyros et al. [2004].

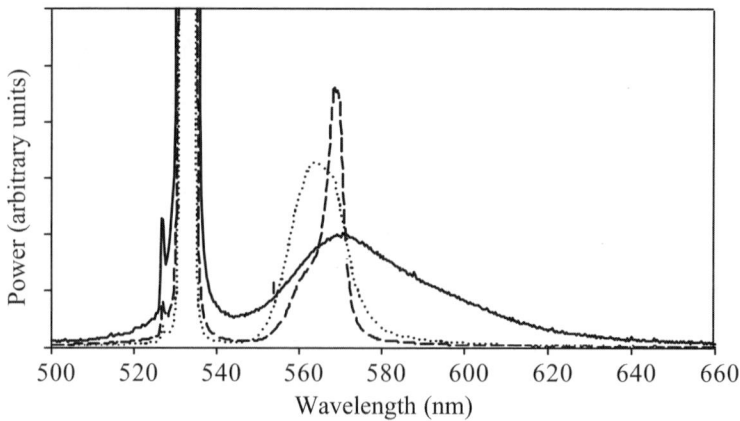

Fig. 11.12. Output of a 50 μmcore R6G-doped mPOF when pumped at 532 nm, showing fluorescence (solid curve), amplified spontaneous emission (dotted curve) and lasing with a 5 nm linewidth (dashed curve). Note that the three curves are not to the same scale, the latter being at much higher power.

To demonstrate the fibre laser a 5 cm long doped fibre with a 100 μmcore diameter was used (shown in Fig. 11.2 of Sec. 11.2.1). Lasing was observed at 568 nm with a linewidth of 5 nm as shown in Fig. 11.12, which is similar to previous reports for PMMA/R6G lasers [Kuriki et al. 2000, Abedin et al. 2003]. Here no external mirrors were used, the cavity relied on Fresnel reflections off the fibre end faces which were not polished. The fibre of Fig. 11.9 used in the amplifier measurements was not used as a fibre laser due to the smaller core size (10 μm) leading to the onset of stimulated Raman scattering which depleted the pump and prevented lasing [Argyros et al. 2005].

11.5 Challenges And Future Directions

Doping is currently the least developed of the applications discussed in this book, but offers many new directions for mPOF research. Polymers can incorporate a large range of dopants, allowing such applications as lasers, improved nonlinear response, and better photosensitivity. These applications have already been explored in POF to some degree, but mPOF offer new opportunities by allowing these performance enhancements to be coupled to those of the microstructure. In silica microstructured fibres this has already lead to better pumping geometries for fibre lasers, for example.

The incorporation of nanoparticles extends these possibilities enormously. As discussed in this chapter, dyes that photo degrade can be isolated in a non-reactive environment, but this is only the most obvious of the potential applications. The incorporation of quantum dots, or metallic nanoparticles and materials for the magneto-otic effect greatly extend these possibilities. The initial experiments, in which nanoparticles are deposited in layers on the surface, need to be improved so that the bulk doped materials can be more used. This inevitably requires the incorporation of the dopants or particles at the polymerisation stage. While there are issues to be address, such as avoiding clustering, and controlling scattering, these experiments represent some of the most exciting prospects for this technology.

References

Abedin, K M, Álvarez, M, Costela, A, García-Moreno, I, García, O, Sastre, R, Coutts, D W, and Webb, C E (2003). 10 kHz repetition rate solid-state dye laser pumped by diode-pumped solid-state laser. *Optics Communications*, 218(4-6):359–63.

Argyros, A, van Eijkelenborg, M A, Jackson, S D, and Mildren, R P (2004). A microstructured polymer fiber laser. *Optics Letters*, 29(16):1882–4. see also [Argyros et al. 2005].

Argyros, A, van Eijkelenborg, M A, Jackson, S D, and Mildren, R P (2005). Reply to comment on "Microstructured polymer fiber laser". *Optics Letters*, 30(14):1829–30.

Barbé, C, Bartlett, J, Kong, L, Finnie, K, Lin, H Q, Larkin, M, Calleja, S, Bush, A, and Calleja, G (2004). Silica particles: A novel drug-delivery system. *Advanced Materials*, 16(21):1959–65.

Barbé, C J A and Bartlett, J (2001). WO Patent 01/62232.

Daum, W, Krauser, J, Zamzow, P E, and Ziemann, O (2002). *POF Polymer Optical Fibers for Data Communication*. Springer Verlag, Berlin, Germany, first edition.

Digonnet, M J F (2001). Continuous-wave silica fibre lasers. In Thompson, B J, editor, *Rare Earth Doped Fiber Lasers and Amplifiers*. Eastern Hemisphere Distribution, Marcel Dekker Inc, The Netherlands.

Finnie, K, Kong, L, Jacques, D, Lin, H-Q, McNiven, S, Calleja, S, Gorissen, E, and Barbé, C (2005). Encapsulation and controlled release from silica particles. *Chemistry in Australia*, 72(2):13–5.

Ishigure, T, Koike, Y, and Fleming, J W (2000). Optimum index profile of the perfluorinated polymer-based GI polymer optical fiber and its dispersion properties. *Journal of Lightwave Technology*, 18(2):178–84.

Kurian, A, George, N A, Paul, B, Nampoori, V P N, and Vallabhan, C P G (2002). Studies on fluorescence efficiency and photodegradation of rhodamine 6G doped PMMA using a dual beam thermal lens thechnique. *Laser Chemistry*, 20(2-4):99–110.

Kuriki, K, Kobayashi, T, Imai, N, Tamura, T, Koike, Y, and Okamoto, Y (2000). Organic dye-doped polymer optical fiber lasers. *Polymers for Advanced Technologies*, 11:612–6.

Kuriki, K, Nishihara, S, Nishizawa, Y, Tagaya, A, Okamoto, Y, and Koike, Y (2001). Fabrication and optical properties of neodymium-, praseodymium- and erbium-chelates-doped optical fibres. *Electronics Letters*, 37(7):415–7.

Large, M C J, Ponrathnam, S, Argyros, A, Pujari, N S, and Cox, F (2004). Solution doping of microstructured polymer optical fibres. *Optics Express*, 12(9):1966–71.

Meissner, K E, Holton, C, and Jr, W B Spillman (2005). Optical characterization of quantum dots entrained in microstructured optical fibers. *Physica E*, 26(1-4):377–81.

Muto, J and Tahjika, S (1986). Diffusion of alcohol-rhodamine 6G in polymethyl methacrylate. *Journal of Materials Science*, 21(6):2114–18.

Naritomi, M, Murofushi, H, and Nakashima, N (2004). Dopants for a perfluorinated graded index polymer optical fiber. *Bulletin of the Chemical Society of Japan*, 77(11):2121–7.

Neogi, P (1996). *Diffusion in Polymers*. Marcel Dekker Inc., New York, USA.

Pang, L, Shen, Y, Tetz, K, and Fainman, Y (2005). PMMA quantum dots composites fabricated via use of pre-polymerization. *Optics Express*, 13(1):44–9.

Shin, B-G, Park, J-H, and Kim, J-J (2004). Plastic photonic crystal fiber fabricated by centrifugal deposition method. *Journal of Nonlinear Optical Physics and Materials*, 13(3-4):519–23.

Song, L, Hennink, E J, Young, I T, and Tanke, H J (1995). Photobleaching kinetics of fluorescein in quantitative fluorescence microscopy. *Biophysical Journal*, 68(6):2588–600.

Tan, W, Wang, K, He, X, Zhao, X J, Drake, T, Wang, L, and Bagwe, R P (2004). Bionanotechnology based on silica nanoparticles. *Medicinal Research Reviews*, 24(5):621–38.

Thomas, N L and Windle, A H (1978). Transport of methanol in poly(methyl methacrylate). *Polymer*, 19(3):255–65.

Thomas, N L and Windle, A H (1980). A deformation model for case II diffusion. *Polymer*, 21(6):613–9.

Yu, H C Y, Barbe, C, Finnie, K, Ladouceur, F, Ng, D, and van Eijkelenborg, M A (2006). Fluorescence from nano-particle doped optical fibres. *Electronics Letters*, 42(11).

Zagari, J, Argyros, A, Barton, G W, Henry, G, Large, M C J, Issa, N A, Poladian, L, and van Eijkelenborg, M A (2004a). Erratum on "small-core single-mode microstructured polymer optical fibre with large external diameter". *Optics Letters*, 29(13):1560.

Zagari, J, Argyros, A, Barton, G W, Henry, G, Large, M C J, Issa, N A, Poladian, L, and van Eijkelenborg, M A (2004b). Small-core single-mode microstructured polymer optical fibre with large external diameter. *Optics Letters*, 29(8):818–20. See also [Zagari et al. 2004a].

Index

Printed in the United States of America.

7
3 R10-E